Applied Crop Physiology: Understanding the Fundamentals of Grain Crop Management

Applied Crop Physiology: Understanding the Fundamentals of Grain Crop Management

Dennis B. Egli
University of Kentucky
USA

CABI

CABI is a trading name of CAB International

CABI
Nosworthy Way
Wallingford
Oxfordshire OX10 8DE
UK

Tel: +44 (0)1491 832111
Fax: +44 (0)1491 833508
E-mail: info@cabi.org
Website: www.cabi.org

CABI
WeWork
One Lincoln St
24th Floor
Boston, MA 02111
USA

Tel: +1 (617)682-9015
E-mail: cabi-nao@cabi.org

A catalogue record for this book is available from the British Library, London, UK.

Library of Congress Cataloging-in-Publication Data

Names: Egli, Dennis B, author.
Title: Applied crop physiology : understanding the fundamentals of grain crop management / Dennis B Egli
Other titles: Understanding the fundamentals of grain crop management
Description: Boston : CAB International, [2021] | Includes bibliographical references and index. | Summary: "This book describes the fundamental processes involved in the accumulation of biomass and the production of grain yield by agronomic crops and discusses how these processes underlie and influence management decisions"-- Provided by publisher.
Identifiers: LCCN 2021018810 (print) | LCCN 2021018811 (ebook) | ISBN 9781789245950 (hardback) | ISBN 9781789245967 (ebook) | ISBN 9781789245974 (epub)
Subjects: LCSH: Crops--Physiology. | Plant physiology. | Plants, Cultivated.
Classification: LCC SB112.5 E35 2021 (print) | LCC SB112.5 (ebook) | DDC 631.5/82--dc23
LC record available at https://lccn.loc.gov/2021018810
LC ebook record available at https://lccn.loc.gov/2021018811

References to Internet websites (URLs) were accurate at the time of writing.

ISBN-13   9781789245950 (hardback)
          9781789245967 (ePDF)
          9781789245974 (ePub)

DOI: 10.1079/9781789245950.0000

Commissioning Editor: Rebecca Stubbs
Editorial Assistant: Emma McCann
Production Editor: Shankari Wilford

Typeset by SPi, Pondicherry, India

Contents

Preface

Books are carriers of civilization, without books, history is silent, literature dumb, science crippled, thought and speculation are at a standstill.

    Henry David Thoreau, American essayist, poet and philosopher (1817–1862)

The trek from the field to the dinner table in modern high-input agricultural systems is long and complicated. The basic raw materials (primarily seeds or grains) must be produced, stored, transported, processed, manipulated and combined with other ingredients before they are presented to the consumer. Losses in harvesting, processing, storage, transport and marketing reduce the quantity that arrives on the consumer's plate, where only a portion is consumed, resulting in additional waste. Functioning of this unbelievably complex system depends upon the availability of technology, social and economic conditions, government policy, international trade and the weather.

This book focuses on only one aspect of this system: the management of grain crop production. This may seem to many, especially non-agriculturalists, to be just one relatively insignificant component of a system with many parts. In reality, production rules, the system cannot function without raw materials; without a constant supply of grains, the rest of the system grinds to a halt.

This book is about grain crops, green plants grown for their seeds; the crops that provide, directly or indirectly, some two-thirds or more of the calories we consume. Fruits and vegetables are important for a healthy diet, but we will ignore them in this book. The production systems of crops grown for their leaves, stems, roots, tubers and tree fruits are so diverse and specialized that they have almost nothing in common with grain crops. There are 20 grain crop species that are widely grown, their average global production in 2016 to 2019 was roughly 3.5 billion tonnes per year (FAOSTAT, 2020). We will not deal individually with all 20 species; maize and soybean will receive more attention because they rank in the top four in global production and because I have more experience with these crops. However, the basic principles of yield production apply to all grain crops, only the details differ.

Those farmers who started planting and harvesting crops some 10,000 years ago were probably just as interested in new production practices that made their life easier and increased their yields as modern farmers are. Their search for improvement was surely a trial-and-error process – a new idea was tested in the field and, if it produced higher yield, it was adopted. In modern times, it would be tested in replicated field experiments at several locations and, if the yield increase was statistically significant, the practice would be recommended. The question of why or how it worked was rarely considered. It is the thesis of this book that a greater focus on why it worked would lead to more productive and more efficient cropping systems and eliminate much of the repeated testing of the same concepts. I believe that knowledge of the fundamental processes responsible for plant growth and the accumulation of yield (the why and how behind crop management decisions) simplifies the decision-making process and helps producers successfully deal with this complex and often chaotic system. Understanding why and how will become even more important as grain producers adjust to new environments created by climate change.

The purpose of this book is to present a simple, straightforward discussion of the principles and processes involved in the production of yield by grain crops. These principles provide the fundamental basis of the management practices used today and the development of new management schemes for the future. This book is intended for the practitioners of agronomic crop production – producers, consultants, agri-business personnel, extension personnel at land-grant universities and scientists involved in applied management research.

Photosynthesis drives growth of all grain crops and economic yield is always the seed, providing the common denominator needed to develop a framework of yield production that applies to all grain crop species. Obviously, details are highly crop specific, but this book is not about details, it is about the basic principles underlying those details. This book does not deal with specific management recommendations; rather its focus is on the principles behind the recommendations. It is my hope that readers of this book will be able to make better, more informed management decisions; decisions that will help maintain a well-fed world in the future.

Acknowledgements

This book grew out of a graduate course (PLS 602, Principles of Yield Physiology) that I taught at the University of Kentucky every other year from 1988 through 2013. The objective of the course was to distil the yield production process in grain crops down to its basic principles, the principles that govern the response of grain crops to management and the environment. This book describes those principles and discusses their involvement in management decisions.

This book is for students interested in grain production, producers and their advisors including county agents, consultants, extension personnel and industry representatives. It is my thesis that an understanding of the principles of yield production will make managers better and better managers will have higher yields and greater profits. This book is not written as a scholarly treatise for other scholars, it is for the practitioners, but it is firmly grounded in the crop physiology literature and in my experiences during a 49-year research career at the University of Kentucky. The general principles in the book apply to all grain crops, but the book is biased towards maize, soybean and Midwestern USA agriculture.

I thank the many students who passed through my course; their questions and comments definitely improved it. I owe a debt of gratitude to Dr Rebecca McCulley, Chair of the Department of Plant and Soil Sciences at the University of Kentucky, for allowing me to keep my office after I retired, and to my colleagues in the department for their support and interactions over the years. A special thanks to Chad Lee, Grains Extension Specialist at the University of Kentucky, Carrie Knott, Grains Extension Specialist at the University of Kentucky Research and Education Center in Princeton, Kentucky, and Steve Crafts-Brandner, my long-time friend and valued colleague, for reviewing sections of this book. I also appreciate Chad's willingness to answer my many questions about the current state of crop management. Their comments were very helpful, but the opinions expressed in this book are solely the responsibility of the author.

Dennis B. Egli
21 January 2021
Lexington, Kentucky

Introduction

1

There are, in fact, two things: science and opinion. The former begats knowledge, the latter ignorance.

Hippocrates (460–370 BC)

Crop Management – The Foundation of Production Agriculture

Most of our food supply comes, directly or indirectly, from seeds harvested from green plants. Our very existence depends on adequate supplies of these seeds, which is determined, in part, by the ability of producers to manage their crops efficiently and sustainably while maximizing productivity. Management decisions can make the difference between crop failure and financial ruin and a record crop and financial success.

The need to manage crops, to select appropriate cropping practices, probably began some 10,000 years ago soon after humankind shifted from living off plants and animals growing in the wild (hunters and gatherers) to planting and harvesting specific plant species, i.e. they became farmers. Why the shift was made is not clear and Diamond (1987) argued that it was a big mistake that ruined our health and contributed to the development of class divisions in society. Whether it was good or not, the change was made and humankind gradually became more dependent on this cycle of planting, nurturing and harvesting and less dependent on what nature could provide in the wild. This change had a tremendous effect on society; it started the shift from everyone being involved in food production for survival to where we are today, with only a small fraction of the population producing food while the rest are freed for other pursuits.

As soon as humans decided to start farming and give up hunting and gathering there was a need for crop management. One can easily imagine that the first farmerers took their management cues from nature; their cropping systems probably mimicked the growth of the plant species they utilized as hunters and gatherers. From this humble beginning, crop management became more

© D.B. Egli 2021. *Applied Crop Physiology: Understanding the Fundamentals of Grain Crop Management* (D.B. Egli)
DOI: 10.1079/9781789245950.0001

complex as the number of choices and decisions increased. The use of more crops, the development of crop rotations, realization of the importance of soil fertility, the advent of multiple varieties of a single crop and the use of herbicides and pesticides to control weeds, insects and diseases steadily increased the complexity of cropping systems and the complexity of the management decisions needed to maximize productivity and economic returns.

The monstrous computer-linked machines and large quantities of off-farm inputs, symbolic of the complexity of modern grain production systems, are a far cry from the production systems used 10,000 years ago at the dawn of agriculture. The basic principles, however, have not changed. A seed is placed in the soil so it can germinate and produce a plant that survives to maturity when it can be harvested. Granted, 10,000 years ago the farmer probably poked a hole in the soil for the seed, while today's farmer rolls across the field in air-conditioned comfort, planting three million soybean seeds per hour with a planter that is monitored by computers and steered by Global Positioning System (GPS). The original farmer probably cut the plant when it was mature, whacked it or stomped on it to knock the seeds loose and tossed the grain into the air to let the wind blow the chaff away. Today's modern combine does exactly the same thing albeit on a monster scale in mechanized glory. The methods have changed drastically, but the process has remained the same for 10,000 years.

There are now signs that basic production process may be changing. Growing 'meat' from animal cells in a nutrient broth is being tested by scientists and, if implemented, will represent a significant change in the way we obtain our protein (meat). Perhaps now the 1932 prediction by the great English statesman Winston Churchill, that 'fifty years hence, we shall escape the absurdity of growing a whole chicken in order to eat the breast or the wing, by growing these parts separately in a suitable medium', will come true, although if it does, science will not have moved as fast as he predicted.

Although the basic processes are the same, the technology used to produce grain crops in the high-input era (starting in the 1930s and 1940s) is constantly changing, forcing producers to adjust their cropping systems to maintain economic viability while maximizing productivity. Management before the high-input era was relatively simple for farmers in the maize belt in the Midwestern USA. They probably followed a standard rotation of maize, small grains and hay, one that their fathers used; they saved their own planting seed, legumes in the rotation along with animal manure provided fertilizer and weeds were controlled mechanically. You could be a successful farmer by just following your father's management system.

Modern farmers, in comparison, face a virtual tidal wave of choices. Selecting varieties from the hundreds available, choosing a tillage system, row spacing, herbicide programme, how much and what kind of fertilizer to apply and what pesticides are needed are examples of the massive increase in complexity in the era of high-input agriculture. The modern farmer cannot possibly be successful using the practices he learned from his father; in fact, some of the practices used just 20 years ago are probably out of date. In addition, a producer's management decisions are made in the face of constantly changing weather and economic (cost of inputs, value of the crop) conditions, some of which are not known when the decision must be made. Good management probably determined which farmer was successful in both systems, but it is much more difficult to be a good manager today than it was 100 years ago.

Seeds Feed the World

The focus of this book is on the management of grain crops: crops where the seed represents economic yield. Grain crops are certainly not the only plant species that feed us; Harlan's (1992) shortlist of cultivated plants used for food contained 352 species. We all appreciate the value of veggies, nuts, tubers and animal products in a healthy diet, but grains are our primary source of calories. Some 60 to 65% of the calories we consume come directly or indirectly (via animals that feed on grain) from just four grain crops: maize, rice, wheat and soybean. These four crops dominate world grain production, representing 85% of the global production of the top 20 grain crops (average of 2016–2019, Table 1.1).

Nine of the top 20 grain crops, including three of the big four, are grasses (*Poaceae*, cereals) and they account for 83% of the global production (average of 2016–2019) shown in Table 1.1. Most of the maize production (no. 1 crop) and some of the other grasses are used for animal feed, for biofuel production (in recent years some 38% of the US maize crop was used to produce ethanol) or for other industrial uses, reducing the food calories available for humans. The use of maize for ethanol production is a relatively recent phenomenon, increasing rapidly after 2000 as interest in using biofuels to combat climate change increased. The impact of this diversion on the food supply is reduced, however, by the use of some of the by-products from ethanol production (distiller's grains) as animal feed. These diversions do not detract from the fact that grasses truly feed the world.

The eight legume crops (*Fabaceae*) in Table 1.1 accounted for only 13% of the total production of the top 20 grain crops, but they are a valuable source of protein. The three crops rounding out the top 20 are two important oil crops, rapeseed (oilseed rape and canola) and sunflower, together accounting for approximately 4% of the top 20 total production, and sesame making only a minuscule contribution (0.2%) to the total production (Table 1.1). The relative importance of these crops will vary by country, but this variation does not diminish the importance of the big four (maize, rice, wheat and soybean) as a source of food.

Seeds of the cereals (grasses) contain high levels of carbohydrates (mostly as starch) (Table 1.1) and only modest concentrations of oil and protein. The legumes, in contrast, have higher levels of protein, making them an excellent complement to the grasses and earning them the title of 'poor man's meat'

Table 1.1. World production and typical seed composition of important grain crops. (Production data from FAOSTAT, 2020; adapted from Egli, 2017.)

Crop		World production[a] (Mt)	Seed composition[b]		
			Carbohydrate $(g\,kg^{-1})$	Oil $(g\,kg^{-1})$	Protein $(g\,kg^{-1})$
Poaceae					
Maize	*Zea mays* L.	1134.80	800	50	100
Wheat	*Triticum* spp.[i]	754.99	750	20	120
Rice	*Oryza sativa* L.	752.39	880	20	80
Barley	*Hordeum vulgare* L.[j]	148.27	760	30	120
Sorghum	*Sorghum bicolor* (L.) Moench	59.95	820	40	120
Millet[c]	*Panicum miliaceum* L.	29.12	690	50	110
Oat	*Avena sativa* L.	23.58	660	80	130
Triticale	× *Triticosecale* Wittmack	14.17	599	18	131
Rye	*Secale cereale* L.	12.37	760	20	120
Fabaceae					
Soybean	*Glycine max* (L.) Merrill	343.44	260	170	370
Groundnut (peanut)[d]	*Arachis hypogaea* L.	48.35	120	480	310
Bean[e]	*Phaseolus vulgaris* L.	29.82	620	20	240
Pea, dry[f]	*Pisum sativum* L.	14.72	520	60	50
Chickpea	*Cicer arietinum* L.	14.21	680	50	230
Cowpea	*Vigna unguiculata* (L.) Walp	8.62	570	10	250
Lentil	*Lens culinaris* Medikus	6.48	670	10	280
Pigeon pea	*Cajanus cajan* L. Millsp.	5.07	560	20	250
Others[g]					
Rapeseed[h]	*Brassica napus* L., *Brassica campestris* L.	72.62	190	480	210
Sunflower	*Helianthus annuus* L.	51.02	480	290	200
Sesame	*Sesamum indicum* L.	5.95	190	540	200

[a]Average of 2016–2019, in megatonnes (millions of metric tonnes) (1 metric tonne (t) = 1000 kg = 2205 lb) (FAOSTAT, 2020).
[b]Seed composition data from Bewley *et al.* (2013), Egli (2017), Hulse *et al.* (1980), Langer and Hill (1991) and Sinclair and de Wit (1975).
[c]May include members of other genera such as *Pennisetum*, *Paspalum*, *Setaria* and *Echinochloa*.
[d]Groundnut (peanut), in the shell.
[e]Also includes other species of *Phaseolus* and, in some countries, *Vigna* species.
[f]May include field pea, *Pisum arvense*.
[g]Rapeseed is in the *Brassicaceae*, sunflower is in the *Asteraceae* and sesame is in the *Pedaliaceae*.
[h]May include industrial and edible (canola) types; data from some countries include mustard (*Brassica juncea* (L.) Czern et Coss).
[i]*Triticum aestivum* L., bread wheat, most common.
[j]Harvested grain usually includes lemma and palea.

(Heiser, 1973, p. 116). Carbohydrate and oil levels in legume seeds vary substantially among species. Soybean and groundnut (peanut) (Table 1.1) stand out from the others with their relatively high oil and protein concentrations; in fact, the oil concentration of groundnut is similar to that of traditional oil crops (oilseed rape, sunflower and sesame).

Selection from an enormous number of grain crop species over the millennia produced the species that provide much of our food today, but there are continuing efforts to find new species to reduce our reliance on just a few grain crops. New crop development is not impossible; soybean and canola (oilseed rape) were new crops in the relatively recent past and today they are very successful mainstream crops. Other attempts, however, have not been very successful. Perhaps there are no more superior crops waiting to be discovered. Maize, rice and wheat were the basis of most important early civilizations (Heiser, 1973, p. 68) and they continue to serve us well.

Grain crops are certainly not our only source of food, but the focus of this book is grain crops. My justification for this focus is threefold. First, grain crops make a substantial contribution to our food supply. We cannot live on lettuce, kale and rocket (arugula). Second, the fundamental basic principles of crop physiology that describe the production of yield are the same for all grain crops, but they may not apply to other non-seed food crops. Each grain crop species will have some unique characteristics that separate it from the others, but collectively they also have many more processes and characteristics in common. This uniformity makes it possible to develop concepts describing the production of yield that apply to all grain crop species. This general approach would be difficult if we included, for example, root crops (e.g. cassava or potato) or leafy vegetables (e.g. lettuce, spinach). Third and finally, my experience is with grain crops. In fact, to make this book manageable and to stay within my area of expertise, the book will feature two crops – maize and soybean – but it will usually be possible to generalize to other grain crops. My 50-some years of research experience will give this book a definite tilt towards the agriculture of the Midwestern USA – the maize and soybean belt.

A Brief History of Crop Productivity

Total production of a crop is determined by yield (weight of seeds, in our case, per unit area) and the harvested area, which is a function of available land resources – land area with climate, soils and topography suitable for crop production – and economic and social conditions. The land area available for grain production is limited and the best land is probably already in use, so expanding the production area may result in lower yields and negative environmental consequences (e.g. clearing forests, increased soil erosion). The effective production area, however, can be increased by growing more than one crop per unit area per year in climates with longer growing seasons. Planting soybean after harvesting a winter wheat crop (double cropping), a common practice in the mid-south in the USA, essentially doubles the harvested area in a year.

The area component of grain production was important in many historical increases in the grain supply. For example, movement of European settlers into the Midwestern USA in the late 1800s (Olmstead and Rhode, 2008, p. 22) and the development of grain crop agriculture in the Cerrado region of Brazil (Caruso, 1997) substantially increased the land area devoted to grain crop production. The shift from animal power (horses and mules) to mechanical power (tractors, trucks fuelled by petroleum) in the early years of the 20th century reduced the land needed for feed production, making more available for food production (Gardner, 2002, p. 12). The contribution of increasing area to higher production levels declined in recent times as the area left for expansion decreased. Interestingly, increasing temperatures and longer growing seasons at higher latitudes associated with climate change may make more land area available for grain production. Changes in rainfall amounts and patterns, on the other hand, could reduce the land available for successful rain-fed production. The complexities that determine the land area available for grain production are well beyond the purview of crop physiologists, so we will not consider this important aspect of the food production system.

Estimates of ancient yields illustrate the dramatic increase since the beginning of agriculture. The yield of maize in 3000 BC in Mexico, estimated from the size of cobs in archaeological excavations, was 100 kg ha^{-1} (about 1.6 bu acre^{-1}) while brown rice yields in Japan in AD 800 were 1000 kg ha^{-1} (893 lb acre^{-1}) (Evans, 1993, pp. 276–279). Wheat yield in England was 500 kg ha^{-1} (7.4 bu acre^{-1}) in AD 1200–1400, but it increased substantially to 1100 kg ha^{-1} (16.4 bu acre^{-1}) by the 1700s and to nearly 2000 kg ha^{-1} (29.8 bu acre^{-1}) by the 1800s (Stanhill, 1976). By comparison, wheat yields in the USA in 1866 were only 740 kg ha^{-1} (11 bu acre^{-1}) (NASS, 2020). This comparison is probably unfair because England's moist, relatively cool climate is better suited for wheat than the drier, warmer climates of much of the US wheat belt. Considering these yields, it is perhaps not surprising that yield in those early years was often expressed as a proportion of the seeding rate. The growth of yield from the beginnings of agriculture until the present is truly extraordinary.

There was no change in US wheat and maize yield from 1866 through 1930 (soybean yields were not estimated before 1924) (Fig. 1.1). Agriculture in the USA (and much of the rest of the world) during this period (1866 to c.1930) was a low-input system that could surely be classified as sustainable and would probably meet today's standards for organic agriculture. Cropping systems in the maize belt in the Midwestern USA were based on rotations involving maize, small grains and forage crops (soybean was not grown for grain until the early 1900s) and an absence of inorganic fertilizer use (Egli, 2008a). Most farms included some form of animal husbandry, so animal manure and forage legumes in the rotation provided organic N for the grain crops. Chemical weed control did not exist, so weeds were controlled by mechanical cultivation, which made it necessary to grow crops in relatively wide rows (~1 m or 40 in wide).

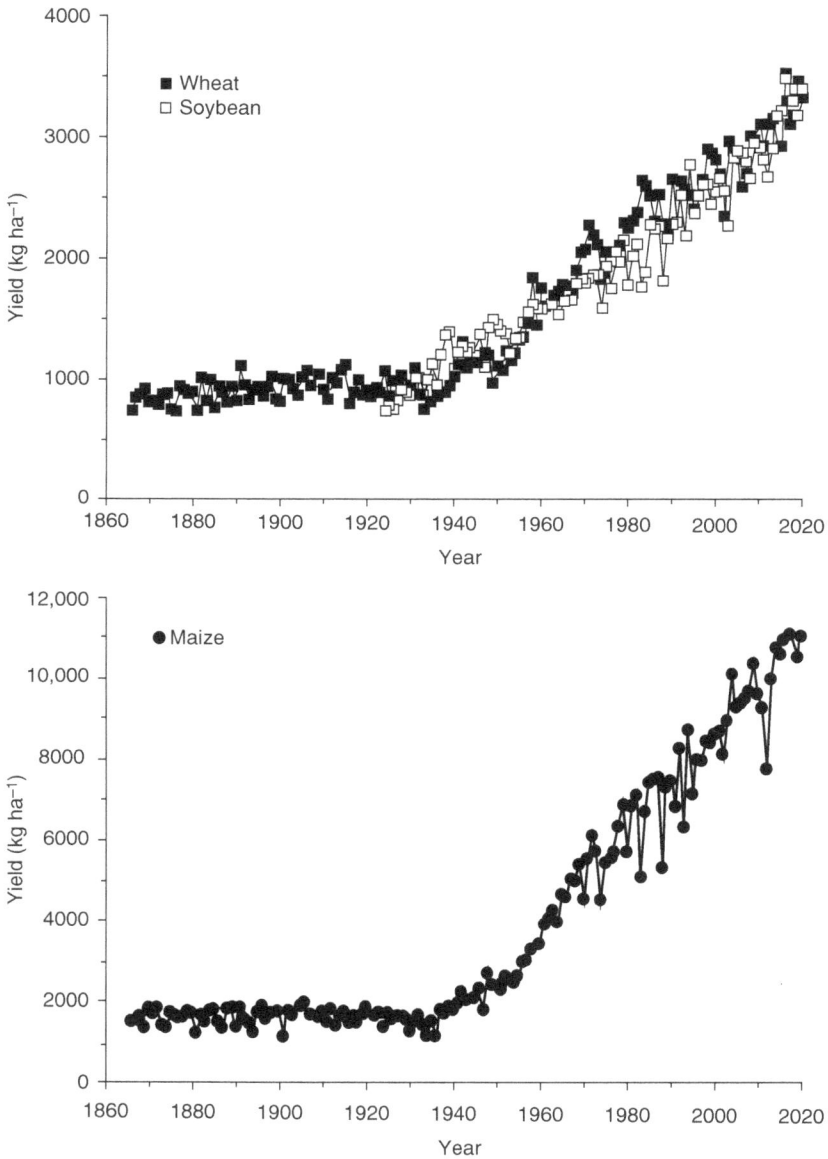

Fig. 1.1. Average yield of wheat, soybean and maize in the USA from 1866 to 2020. The average yield of wheat from 1866 to 1930 was 912 kg ha^{-1} (13.6 bu acre^{-1}); the average of maize over the same period was 1638 kg ha^{-1} (26.1 bu acre^{-1}). (Data from NASS, 2020.)

Farms were small (~20 ha or 50 acres) and farmers grew open-pollinated maize and generally saved their own seed for next year's planting seed. State extension specialists from land-grant agricultural universities conducted 'corn' schools in states with large maize acreage to teach farmers how to select the perfect ear to save to plant next year's crop. Papers in the *Journal of the American Society of Agronomy* (first published in 1908) from this era describe field research into practical aspects of maize production thought to influence yield. In spite of these efforts, maize yields in the USA did not change until the advent of high-input modern agriculture. Today's grain producers expect constantly increasing yields; they would be shocked by a yield plateau that existed for over a half century. This long-lasting yield plateau was also found in other grain-producing areas of the world.

Agriculture during this era in the Midwestern USA was very similar to that proposed by critics of modern agriculture who favour low-input, organic, sustainable production systems. It is difficult, however, to imagine how these systems, with their relatively high labour requirement, would fit into modern society where less than 2% of the US population is directly engaged in production agriculture (Gardner, 2002, p. 93). The dramatic decline in the proportion of the US workforce involved in agriculture from 40% in 1940 to current levels of less than 2% suggests that agriculture is not a preferred occupation for many people. Reversing this trend may be difficult.

Global yields of wheat, rice and maize have increased steadily since 1960 (Fig. 1.2) with no evidence that they are plateauing. Grain crop yields in the

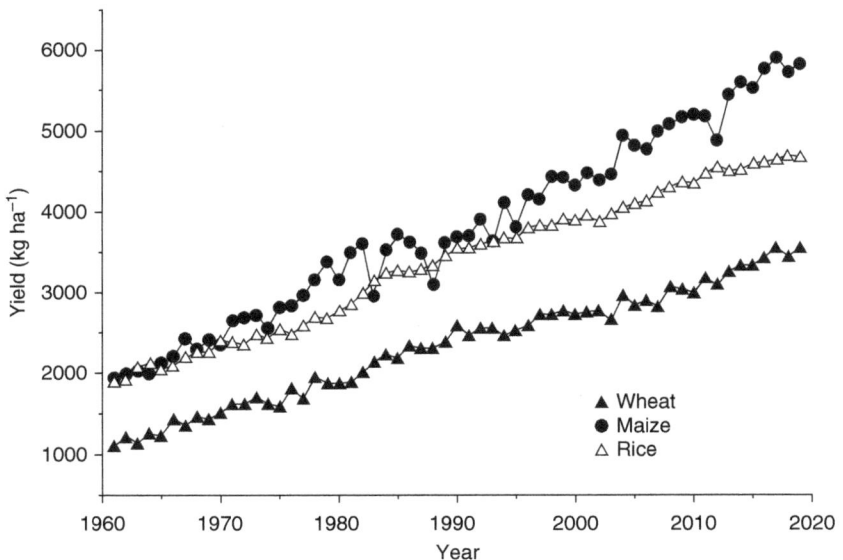

Fig. 1.2. Average world yield of maize, rice and wheat from 1961 to 2019. (Data from FAOSTAT, 2020.)

USA (maize, wheat and soybean) (NASS, 2020) also increased steadily since the 1940s and, in common with global trends (Fig. 1.1), show no evidence that the increase is ending. Yields in some countries, however, have plateaued (Fig. 1.3), but the causes of these plateaus are not clear. They could be a result of changes in government policy or economic conditions that restrict inputs.

Yield growth is often restricted in high-stress, low-yield environments. Non-irrigated soybean yields did not increase from 1972 to 2003 in 45% of the Nebraska counties and 80% of the Arkansas counties evaluated by Egli (2008b). Irrigated yields increased significantly in the same counties. The relative rate of growth (% year^{-1}) of county soybean yields in Kentucky decreased as the proportion of the soybean production area in each county devoted to double cropping after wheat increased (Fig. 1.4). Double cropping after winter wheat necessitates planting soybean after the optimum date, causing a reduction in yield; apparently, the stress of the late planting reduced the rate of yield growth. Although US grain yields increased, on average, since the 1940s, the increase was very much dependent upon the quality of the environment where they were grown.

The dramatic change that ended the 70-year yield plateau in the 1930s was associated with the advent of high-input, so-called 'industrial' agriculture that rapidly replaced the traditional farming systems. The development of improved varieties by plant breeders, including the replacement of open-pollinated maize varieties with hybrids, provided the foundation for the yield growth. The deployment of hybrid maize in the US maize belt began in the 1930s and it was grown on 50% of the area by 1940 and 90% by 1950 (Russell, 1991).

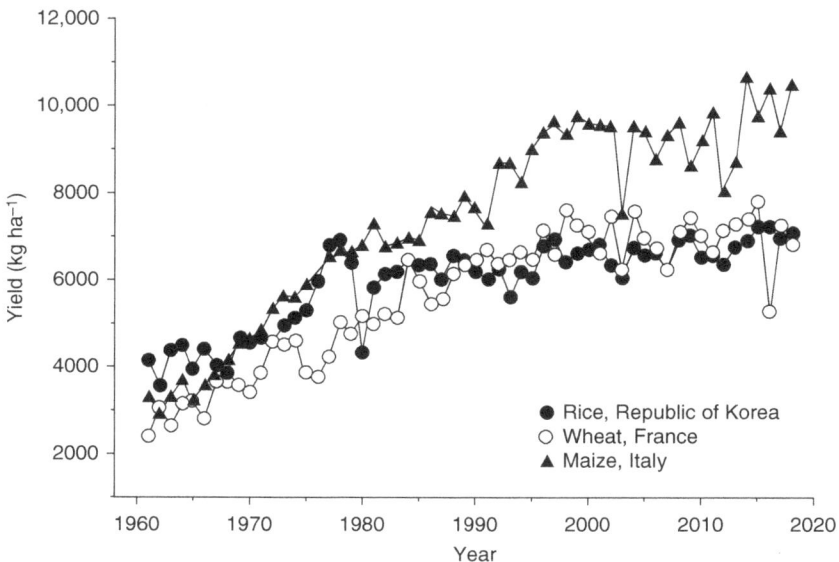

Fig. 1.3. Yield trends of maize, wheat and rice from 1961 to 2018 in countries exhibiting clear yield plateaus. (Data from FAOSTAT, 2020.)

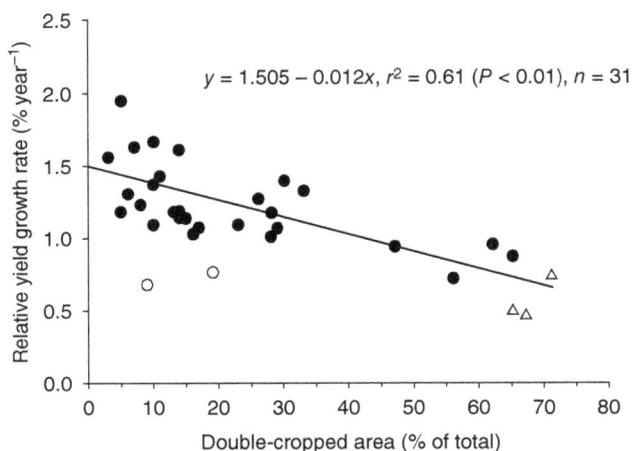

$y = 1.505 - 0.012x$, $r^2 = 0.61$ ($P < 0.01$), $n = 31$

Y-axis: Relative yield growth rate (% year^{-1})

X-axis: Double-cropped area (% of total)

Fig. 1.4. The relationship between the proportion of the soybean area in a county devoted to double-cropped soybean (soybean grown as a second crop after winter wheat) and the relative rate of soybean yield gain (percentage per annum) in 33 counties in Kentucky, USA (1972 to 2003). The area devoted to double-cropped soybean was assumed to equal the harvested wheat area in each county. The open circles (O) and triangles (Δ) represent counties where the yield growth was not significantly different from zero ($P = 0.05$). The open circles were not included in the regression analysis. (From Egli, 2008b.)

The adoption occurred first in the heart of the maize belt (Iowa reached 90% by 1940) followed closely by the surrounding states (hybrids occupied less than 10% of the acreage in Kentucky in 1940, but this increased to 90% by 1950) (Griliches, 1957). The use of inorganic N fertilizer increased rapidly after 1945 (Thompson, 1969), reducing the dependence on animal manures and legumes in the rotation for N. Herbicides for weed control appeared on the scene at this time (~40% of the maize area in Illinois was treated by 1960) (Pike *et al.*, 1991). The continuing trend for mechanization of farming operations probably contributed to the increase in yield by improving the timeliness of critical management operations.

It is interesting that these dramatic yield increases occurred nearly simultaneously in all major grain crop species in spite of significant differences in their physiology, morphology and seed characteristics. Maize produces all of its high-starch seeds on a compact ear in the middle of the plant, it produces high yield with C_4-type photosynthesis (see Chapter 2, this volume) and requires high levels of N fertilizer. Soybean, a legume that produces its own N, has C_3-type photosynthesis and produces seeds with high levels of oil and protein that are evenly distributed over the entire plant. Wheat produces its high-starch seeds in a compact ear at the top of the main stem (and tillers), has C_3-type photosynthesis and responds to N fertilizer. Soybean and wheat varieties are inbred lines, not hybrids like maize. In spite of this diversity, all of the crops responded to high-input agriculture with dramatically increased yields. In fact, the relative rate of yield increase of maize and soybean has been essentially the same since 1980, as shown by a constant ratio of maize yield to soybean yield during that time (average ratio = 3.26) (Fig. 1.5).

The value of the individual components of the new technology was, of course, very crop specific. As noted, improved varieties created by plant breeders provided the foundation of the yield increase in all three crops, but improved varieties cannot produce high yields without adequate fertilizer, weed control and optimization of other aspects of the cropping system. Conversely, old varieties will not produce modern yields with high levels of fertilizer and perfect weed, insect and disease control. Maize and wheat benefited from the widespread use of N fertilizer, but soybean did not. The timing of the beginning of the use of herbicides for weed control varied among crops. Increased maize yields required higher plant populations, but soybean and wheat did not. The yield increase in all crops was driven by improved varieties, but the utilization of specific management practices that removed negative aspects from the environment of each crop (lack of fertilizer, presence of weeds, diseases and insects, failure to intercept all of the incident solar radiation, etc.) was necessary to fully realize the yield potential of improved varieties. In spite of this crop-specific diversity, the relative rates of yield growth for two of the crops, maize and soybean, were the same (Fig. 1.5).

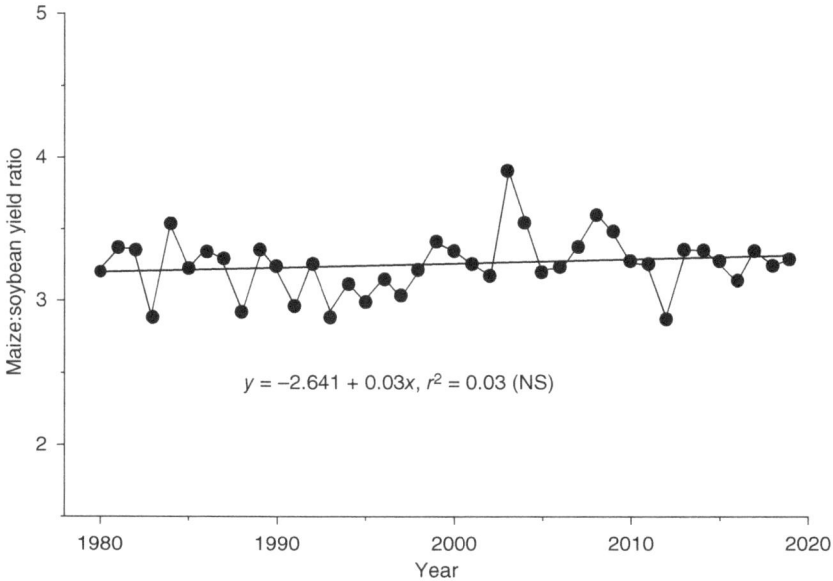

$$y = -2.641 + 0.03x, r^2 = 0.03 \text{ (NS)}$$

Fig. 1.5. The ratio of maize yield to soybean yield in the USA from 1980 to 2019. Yields were converted from $bu\,acre^{-1}$ to $kg\,ha^{-1}$ before calculating the ratio. The average ratio was 3.26. NS, not significant. (Data from NASS, 2020.)

Today's yields of important grain crops are four- to sixfold higher than they were in 1930 (Fig. 1.1), an extraordinary increase in only 90 years, especially in comparison with the 70-year yield plateau that preceded these dramatic increases. These huge increases in productivity are the result of a production system that is much more complicated than those used in the years before 1930. Producers today must make many more decisions than their counterparts of yesteryear. The advent of a much larger array of inputs (fertilizers, crop protection chemicals, herbicides, seed treatments, magic potions, etc.), varieties with genetically modified organism (GMO) traits such as herbicide tolerance and insect resistance, and precision agriculture techniques increased the number of options almost exponentially. This complexity requires greater managerial skills, for producers and the consultants who advise them, to keep the system operating efficiently, both agronomically and economically. The development of big data, sophisticated algorithms and artificial intelligence may help deal with this complexity, although the value of these approaches is yet to be determined.

Modern high-input grain production systems are often plagued by overproduction and low prices, so operating at peak efficiency is critical to a producer's economic survival. I think understanding the physiology of crop production is one key to managing these complex crop production systems to efficiently and sustainably produce maximum economic yield. Successful management of these complex systems requires knowledge of how the heart of the system – the crop community (a field of crop plants) – functions. What are the important processes that determine the productivity of this community? How are they affected by the environment? How can this community be manipulated to increase yield or to improve production efficiency? These questions are central to the maintenance of a profitable grain production system, especially when facing the effects of climate change, and they make up the central theme of this book.

Crop Management and Yield

The primary goal of crop management is to increase crop growth and yield by improving the crop's environment. A second important goal is to improve the efficiency of the system by producing more output from the same level of inputs or the same output from a lower level of inputs. Controlling weeds and diseases, applying fertilizer and irrigating improve the crop's environment by supplying needed inputs or removing negative factors. The optimum planting date positions the growth of the crop in the most favourable environment. Improving the environment by management implies that the producer understands the characteristics of the perfect environment – the environment that maximizes growth and yield – for each crop.

Historically, and still too often today, the search for the perfect environment was a matter of trial and error – a new management practice was tested and, if it worked (higher yield or greater efficiency), it was adopted, if not it was abandoned. The question of why it worked or why it did not work was not considered. Producers, agri-business personnel, and extension and research workers at land-grant universities and the US Department of Agriculture (USDA) are constantly evaluating management practices, searching for the combination that will produce greater efficiency and/or higher yield. The trial-and-error approach often results in nearly constant re-evaluation of the same questions. Without asking why, there is no accumulation of knowledge over time that would eventually make it possible to predict the response. Variety (constantly changing over time) by management interactions are frequently used to justify this constant re-evaluation, but seldom is the existence of a variety by management interaction documented. The first soybean row spacing experiment in the USA that I am aware of was published in 1939 (Wiggans, 1939); 80-some years later we are still experimenting to find the best row spacing (Schmitz *et al.*, 2020). This situation is an embarrassment to the discipline of Crop Physiology and it is a direct result of not focusing on the why question. The underlying thesis of this book is that a greater focus on the why question leads to more productive, efficient and sustainable cropping systems.

The production of yield by a grain crop is a dynamic process expressed over a finite time, and it involves all of the myriad reactions, processes, cycles and systems that make up plant growth. The investigation of these processes, cycles and systems has a long history. The rudimentary aspects of photosynthesis, the process that is the primary driving force behind the production of yield, were elucidated in the late 1700s (King, 1997, pp. 19–20). These observations essentially defined the photosynthetic process; it wasn't until the 1950s and 1960s and the availability of ^{14}C that the carbon fixation cycle was described in detail and the two types (C_3 and C_4) of photosynthesis in crop plants were discovered. de Sussine discovered that plants take up minerals and NO_3^- from the soil in 1804. Boussingault found that legumes could get their N from air in 1837, but the details were not known until the work of Hellriegel and Beijerknak in 1888 (Evans, 1975, p. 2).

From these rudimentary beginnings, we now have a detailed understanding of the physiological processes that produce growth, the environmental conditions and inputs required for growth and how the plant responds to its environment. This knowledge extends from the subcellular level of molecules, enzymes, cells and genetic information to the whole plant and the plant community (e.g. soybean or maize field). Crop physiology, defined by the eminent Australian crop physiologist Lloyd Evans (Evans, 1975, p. 13) as 'understanding the dynamics of yield development in crops', contributed significantly to this understanding. Crop physiology is largely focused on the why question of crop management and it probably traces its origins to the studies of plant spacing and sowing date with cotton in Egypt by W.L. Balls in the early 1900s (Evans, 1975, p. 13). Crop physiologists focus much of their research on the crop community because many of the processes that are important yield determinants are expressed only at the community level. Studying isolated plants or plant parts rarely produces useful information for managing crops, but the basic tenets of crop physiology provide the why and how of crop responses to management. Unfortunately, much of this knowledge of the yield production processes is, even today, ignored in the never-ending search for higher yield and greater efficiency.

Purpose

The purpose of this book is to present a simple, straightforward discussion of the yield production process at the community level to aid the search for higher yield and greater efficiency of production. The yield production process forms the basis of the crop management practices used today and undergirds the search for new management practices. Surely, an understanding of these processes provides a much better basis for the development and evaluation of new management systems than random trial and error. Imagine the characteristics of the automobiles we would be driving today if the automotive engineers of yesteryear designed new improved automobiles by first surveying all the cars in a large parking lot, measuring their characteristics and using this database to select traits that seemed to be associated with higher speed or better gas mileage. Automotive engineers were successful because they understood how automobiles worked and used that knowledge to design better automobiles. We know a lot about how crop communities produce yield; the challenge is to apply this information to improving crop management systems.

We will begin our journey to understand the yield production process by discussing the fundamental growth processes in Chapter 2. First, we will cover photosynthesis and respiration, the two processes directly responsible for the accumulation of weight by plants. Then we will discuss seed growth, the processes by which the seeds utilize raw materials from photosynthesis to produce the storage materials that give them their value. Finally, we will discuss water, the primary yield determinant in most rain-fed agricultural systems. The growth of crop communities will be the focus of Chapter 3, followed in Chapter 4 by the use of the knowledge we gained in Chapters 2 and 3 to understand the fundamentals of the basic management decisions involved in grain crop production. The last chapter, Chapter 5, addresses some of the challenges and opportunities facing agriculture and humankind in the years ahead.

Chapters 1 through 3 provide the knowledge needed to construct a unified model of the yield production process that applies to all grain crops. Crops differ in specific details and management recommendations, but this book is not about details or recommendations, it is about the basic principles underlying those details.

Units of measure are always an issue when writing about agriculture for a global audience. The metric system is utilized throughout this book with non-metric units included for readers not familiar with the metric system. A conversion table (Appendix Table A1) is included to facilitate conversion of units.

The common name of grain crops may vary among locations (are sorghum, kafir corn and milo three crops or one?), but there is only one scientific name for each crop (in the example, there is only one crop, *Sorghum bicolor* (L.) Moench, with three common names). Appendix Table A2 contains the scientific names of all plant species referred to in this book.

Basic Plant Growth Processes

2

Flex que potuit rerun cognosceve causes (Fortunate is he who understands the cause of things).

Virgil, Italian poet (70–19 BC)

Introduction

Crop growth and yield respond to management because the management practices make the environment in the crop community better suited for crop growth. Yield goes up because the crop community and/or the processes making up plant growth respond favourably to the improved environment. Consequently, to understand how and why changes in management practices affect yield, we must consider the fundamental plant processes that collectively are responsible for growth. We will discuss these processes in this chapter.

Understanding these fundamental plant processes at the cellular, organ and plant levels is important, but it is only part of the story. Yield is produced by a crop community, a collection of plants, a field of maize, soybean or rice, not by isolated plants. The expression of yield as weight per unit area, not weight per plant, highlights the importance of the community. Individual plant characteristics are often overshadowed by the characteristics of the crop community, so ultimately crop growth represents a blend of individual plant and community characteristics. We will focus on individual plant processes – photosynthesis, respiration, water use, leaf senescence and seed growth – in this chapter and develop the community relationships that are necessary to understand the production of yield in Chapter 3.

A maize producer plants 25 kg of seed on a hectare (22 lb acre^{-1}) and, 100 to 120 days later, harvests 15,677 kg ha^{-1} (14,000 lb acre^{-1}, 250 bu acre^{-1}) of seed and leaves roughly 15,677 kg ha^{-1} of vegetative material (stover) (not counting the roots) in the field. The weight of the planted seed increased by roughly 600 times in this 3- to 4-month period; this is the miracle of agriculture that feeds the world.

© D.B. Egli 2021. *Applied Crop Physiology: Understanding the Fundamentals of Grain Crop Management* (D.B. Egli)
DOI: 10.1079/9781789245950.0002

This huge increase in weight is the result of a single plant process – photosynthesis – that occurs primarily in the green leaves of the crop community. This miracle occurs in all grain crops and provides us, directly or indirectly, with all of our food. The green leaves use energy from sunlight to take CO_2 from the air and incorporate it into simple sugars that serve as building blocks to construct the many compounds making up a plant. We derive all of our sustenance from the sun via the process of photosynthesis. An adequate supply of mineral nutrients, often from fertilizers, is necessary to ensure the efficient functioning of photosynthesis, but these mineral nutrients contribute only a tiny portion to the weight of the plant.

Photosynthesis does not just provide us with food and fibre, for eons it was and still is our primary source of energy. The wood that heated our homes and cooked our food, the coal that fuelled the industrial revolution and generates much of our electricity, and the petroleum that powers our tractors, automobiles and airplanes are all products of photosynthesis.

Managing grain crops is a matter of managing photosynthesis – creating an environment in our fields of maize, soybean or wheat that will maximize photosynthesis of the crop community. Environmental conditions or management practices that increase yield must result in an increase in the rate or duration of community photosynthesis. Fertilizing, irrigating and controlling weeds, diseases and insects all increase yield because they increase photosynthesis.

Other important processes that will be investigated in this chapter include respiration, the process that captures energy from the breakdown of simple sugars for use in a wide variety of metabolic processes (acquisition of N, synthesis of all the compounds making up the plant, etc.), leaf senescence and water use by crops. These processes are essential components of crop growth, but photosynthesis is the driving force behind the production of yield.

Finally, we will include seed growth in our discussion. The yield of grain crops is the seeds that are harvested at maturity; consequently, we must discuss the capacity of seeds to use the raw materials (basically sucrose and a few amino acids) from the mother plant to grow and produce the storage compounds (oil, protein and starch) that make seeds valuable. Considering only the capacity of the community to produce raw materials via photosynthesis is not enough, we must involve seeds to get a complete picture of the yield production process.

Photosynthesis

Our goal in this section is to understand the basic photosynthesis process, the basis for species differences and responses to the environment. We can achieve this understanding without delving too deeply into the biochemical intricacies of the process.

Our knowledge of the photosynthetic process goes back several hundreds of years to when scientists first pondered how a tiny seed could produce such a large plant. It seemed logical at that time to assume that plants got their sustenance from the soil, but in 1648 Jan van Helmont, a Belgian physician, demonstrated that a willow tree grown in a large pot increased substantially in weight over a 5-year period without a corresponding decrease in the weight of the soil in the pot (King, 1997, p. 18). Based on the knowledge available at that time, he mistakenly concluded that the weight gain came from water. In 1771, Joseph Priestly, an English clergyman and scientist, observed that a candle burning in an airtight container soon went out and a mouse in the container died. He concluded that combustion 'injured' air. Priestly demonstrated that this noxious air could be made breathable by plants growing in the container, i.e. plants released O_2 during growth. Jan Ingenhousz, a Dutch physiologist, demonstrated in 1782 that purification of the air by green plants occurred only in the light. He went on to show that plants take up CO_2. A Swiss scientist, Nicolas-Théodore de Saussure, completed the story in the early 1800s when he demonstrated that the gain in carbon from CO_2 uptake did not account for all of the weight gain by the plant, the rest came from water (King, 1997, pp. 19–20). It was now clear that a small seed could produce a large plant by using energy from sunlight to take CO_2 from the air and convert it into plant tissue. Today we have a very detailed knowledge of the photosynthetic process, all the way from the level of the molecule, enzyme and cycle to the crop community.

Our current understanding of photosynthesis is summarized in Eqn 2.1:

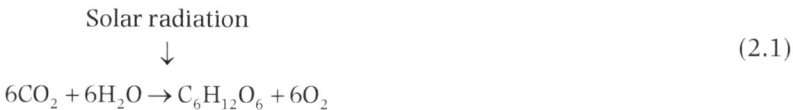

$$\text{Solar radiation}$$
$$\downarrow \qquad\qquad (2.1)$$
$$6CO_2 + 6H_2O \rightarrow C_6H_{12}O_6 + 6O_2$$

Solar radiation, absorbed by chlorophyll in green leaves, provides the energy to drive the production of simple sugars from CO_2 and H_2O resulting in the evolution of O_2. The energy in the absorbed solar radiation splits H_2O producing O_2 and high-energy compounds that provide the energy to incorporate CO_2 into simple sugars which are used as building blocks to produce all of the compounds making up the plant. CO_2 moves from the atmosphere into the leaf through the stomata (pores in the leaf surface) (Fig. 2.1) while water vapour moves out of the leaf through these same pores (transpiration). Consequently, photosynthesis (CO_2 fixation) is inseparably linked to water loss via transpiration. The water used as a reactant in photosynthesis (Eqn 2.1) is only a small portion of the total water use by the crop. Most of the water simply passes through the crop and into the atmosphere. The stomata close when water is limiting to reduce water loss, but, unfortunately, closing the stomata may also limit CO_2 movement into the leaf, thereby reducing photosynthesis and plant growth. This unfortunate linkage between photosynthesis and water loss is the primary reason why availability of water is a major determinant of crop productivity. We will discuss water availability at some length later in this chapter.

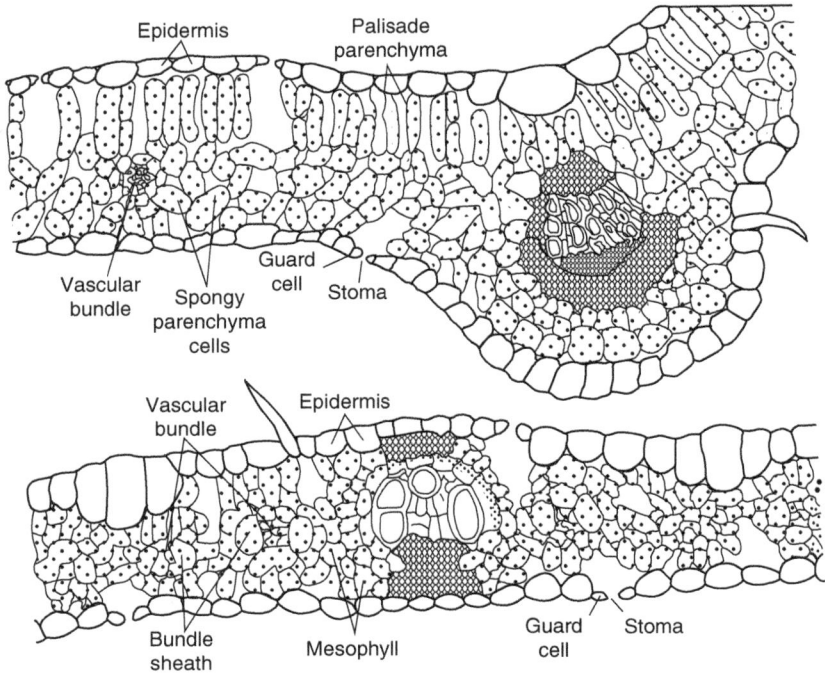

Fig. 2.1. Cross-sections through a leaf from a C_3 species (lucerne (alfalfa)) (top) and a C_4 species (maize) (bottom). CO_2 and water vapour move into and out of the leaf through the stoma (plural: stomata). (Adapted from Gardner *et al.*, 1985, p. 17. Used with permission from John Wiley and Sons.)

C_3 and C_4 photosynthesis

The carbon cycle in photosynthesis was first described by Melvin Calvin and his associates at the University of California at Berkley in the 1950s. They used radioactive CO_2 ($^{14}CO_2$) to follow carbon from CO_2 in the atmosphere to the formation of simple sugars. They found that carbon from CO_2 was added to a five-carbon sugar (ribulose bisphosphate, RuBP) by the enzyme ribulose bisphosphate carboxylase/oxygenase (Rubisco), producing an unstable intermediate that immediately spilt into two molecules of a three-carbon compound (3-phosphoglycerate, 3-PGA). Hence, the designation of the process as C_3 photosynthesis. The 3-PGA is the starting point for the synthesis of all the materials that make up the plant.

In addition to fixing carbon, Rubisco also acts as an oxygenase, catalysing the oxidation of RuBP that subsequently releases CO_2 into the atmosphere. This process is called photorespiration because it occurs only in the light. Photorespiration reduces the amount of carbon fixed by photosynthesis, thereby reducing plant growth. CO_2 and O_2 compete for the same active sites on Rubisco, so photorespiration and the loss of CO_2 decrease as the CO_2 concentration in the air around the leaf increases. Consequently, higher CO_2 concentrations in the atmosphere increase photosynthesis and yield of plant species with C_3 photosynthesis (Fig. 2.2). A summary of field studies showed an increase in soybean yield of 34 to 38% when CO_2 concentration in the air was doubled from approximately 387 ppm (Hatfield *et al.*, 2011). Similar responses occurred for other C_3 crops.

In the early 1960s, two scientists – H.R. Kortschalk (sugarcane) and Y. Karplar (maize) – discovered a new carbon fixation cycle in these species when they were trying to replicate the work of Calvin. The details of this cycle were described by M.D. Hatch and C.R. Slack in Australia; hence this type of photosynthesis is often referred to as Hatch and Slack photosynthesis. They found that CO_2 was fixed in these species by the enzyme phosphoenolpyruvate (PEP) carboxylase, forming a four-carbon compound (malate); hence this system is also known as C_4 photosynthesis. Malate is transported into specialized cells surrounding the vascular tissue (bundle sheath cells) (bottom of Fig. 2.1) where CO_2 is split off malate and re-fixed by Rubisco and the C_3 cycle. The key aspect of this re-fixing is that the CO_2 concentration in the bundle sheath cells is very high, completely inhibiting the oxygenase activity of Rubisco and

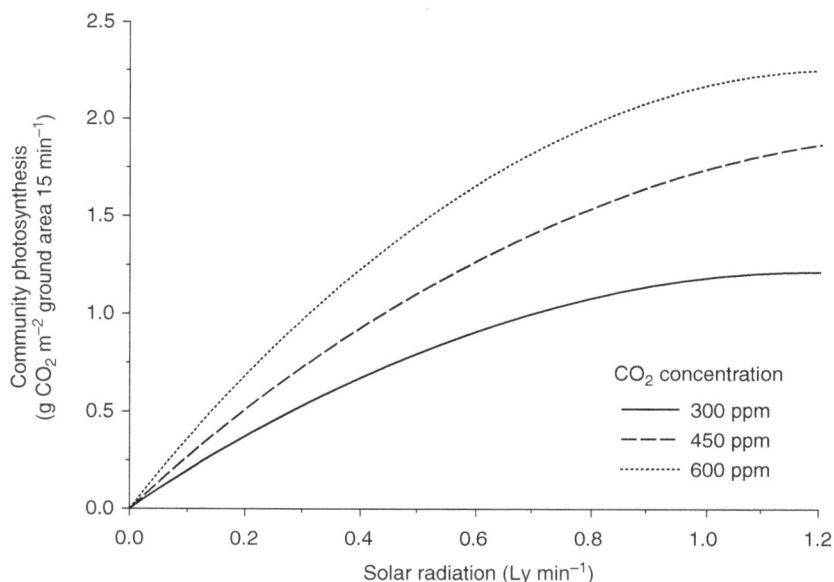

Fig. 2.2. Photosynthesis of a soybean (a C_3 species) community (variety 'Harasoy') in the field as a function of the CO_2 concentration and solar radiation levels. Current atmospheric CO_2 concentration is approximately 414 ppm. Maximum solar radiation levels in mid-summer on a clear day would be approximately 1.2 Ly min^{-1}. Ly, Langley. (Redrawn from Egli *et al.*, 1970.)

reducing photorespiration to zero. Consequently, crop species with C_4 photosynthesis usually have higher rates of photosynthesis and higher potential yields than species with C_3 photosynthesis. Photosynthesis of plants with the C_4 pathway does not increase as the CO_2 concentration in the atmosphere increases above current levels (~414 ppm) because Rubisco is already saturated with CO_2.

In spite of the higher productivity associated with C_4 photosynthesis, most of the major grain crops, including wheat, rice and soybean, have C_3 photosynthesis (Table 2.1). The important root crops, potato, sugarbeet and cassava, are also C_3 species. Maize, sorghum and millet are the only notable grain crops with C_4 photosynthesis. Sugarcane is also an important C_4 crop. It is interesting that the relatively inferior C_3 photosynthesis system is responsible for 65% of the world grain production reported in Table 1.1; the C_3 contribution to the world's food supply is even larger when potato and cassava are included. The higher photosynthetic potential of C_4 species seems to be underutilized. However, scientists are currently working to transfer the C_4 system into rice to increase its productivity (Ermakova *et al.*, 2020), which, if successful, would help resolve the conundrum of increasing food production without increasing the area used to produce food crops.

Photosynthesis and the environment

Environmental conditions (temperature, solar radiation levels, water availability and CO_2 concentrations) during the growing season play an important role in determining yield in a producer's field. Since canopy photosynthesis produces yield, we must investigate the effect of the environment on the rate of

Table 2.1. Leading grain crop species characterized by their photosynthetic system. Species are listed in approximate order of their world production as shown in Table 1.1.

Photosynthesis system		
C_3		C_4
Rice	Triticale	Maize
Wheat	Rye	Sorghum
Soybean	Pea	Millet
Barley	Chickpea	
Sunflower	Cowpea	
Groundnut (peanut)	Lentil	
Rapeseed[a]	Pigeon pea	
Bean	Sesame	
Oat		

[a]Canola and oilseed rape.

photosynthesis. The rate of photosynthesis is tightly linked to the availability of water (as discussed previously), which we will revisit in detail later in this chapter.

It is difficult to measure the rate of photosynthesis of a crop community; so much of the information describing environmental effects on photosynthesis comes from measurements with single leaves, not crop communities. Photosynthesis of a leaf is measured with an instrument that seals a portion of a leaf in a plastic chamber that allows solar radiation to reach the leaf. The instrument measures the decrease in CO_2 concentration of air as it flows through the chamber, providing an estimate of the CO_2 taken up by the leaf, i.e. photosynthesis. Environmental conditions in the chamber are varied to investigate their effect on photosynthesis.

The crop in the field is constantly exposed to varying levels of all of the environmental factors that influence photosynthesis, but this complexity is not necessarily captured by experiments evaluating only a single environmental characteristic in the laboratory. These simple experiments do not capture interactions between factors. Interactions occur when the effect of one factor on photosynthesis depends on the level of other factors; for example, when the effect of solar radiation on the rate of photosynthesis depends on air temperature, the concentration of CO_2 in the atmosphere (Fig. 2.2) or the availability of water. Interactions occur in all aspects of crop growth, not just photosynthesis, and they are often complex and difficult to understand, but they represent the real world where crops are grown and we must learn to deal with them. Understanding the growth of crop communities in the field would benefit from more information on community photosynthesis, but it is not a popular research topic, so we must make do with what we have.

Solar radiation

Since solar radiation drives photosynthesis, it is not surprising that single-leaf photosynthesis ($g\,CO_2\,m^{-2}$ leaf area h^{-1}) increased as solar radiation (irradiance) increased for both C_3 and C_4 crop species (Fig. 2.3). Photosynthesis of the C_3 species reached its maximum rate (~2.5 $g\,CO_2\,m^{-2}\,h^{-1}$) at relatively low solar radiation levels (~25% of maximum) while the C_4 species continued to increase (maximum rate > 7.5 $g\,CO_2\,m^{-2}\,h^{-1}$) to the highest level (which approximates maximum solar radiation on a clear sunny day in mid-summer at mid-latitudes).

The single-leaf curve in Fig. 2.3 predicts that solar radiation could be reduced substantially without affecting photosynthesis and yield of a C_3 species. Photosynthesis of communities of C_3 plants ($g\,CO_2\,m^{-2}$ ground area h^{-1}), however, does not saturate at low solar radiation levels; it continues to increase as solar radiation increases (Fig. 2.2), but at a slower rate than for C_4 species. The upper leaves of the crop canopy are saturated at relatively low levels of solar radiation (see Fig. 2.3), but as solar radiation increases more of it penetrates to lower leaves, which increases their photosynthesis rate and the rate of the community (Fig. 2.2). Artificially reducing solar radiation levels with shade

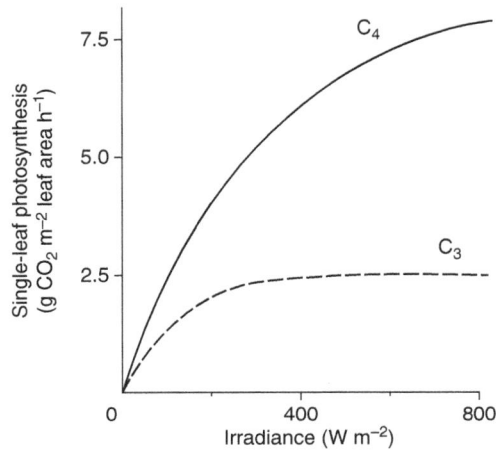

Fig. 2.3. Response of single-leaf photosynthesis of a C_3 and C_4 species to increasing levels of solar radiation (irradiance) at ambient CO_2 levels and optimum temperatures. Solar radiation level on a clear sunny day in mid-summer in the US maize belt is roughly 500 W m^{-2}. (From Conner *et al.*, 2011, p. 268.)

cloth in the field reduced crop growth rate (an estimate of community photosynthesis) and yield of soybean, a C_3 species (Egli and Yu, 1991). The same response will occur in maize, a C_4 species (Gao *et al.*, 2017). Photosynthesis of crop communities in the field is directly dependent upon solar radiation levels, which is not surprising given that solar radiation provides the energy for photosynthesis and plant growth. High yields require high levels of solar radiation.

Carbon dioxide

As discussed previously, increasing atmospheric CO_2 levels increases C_3 photosynthesis (Fig. 2.2) because there are more CO_2 molecules available to compete with O_2 for active sites on Rubisco, thereby reducing oxygenase activity and increasing carboxylase activity, carbon fixation and growth. Photosynthesis of C_4 species reaches maximum levels at ambient CO_2 concentrations and does not respond directly to CO_2 concentrations above ambient. CO_2 levels above ambient cause stomata to partially close, which reduces water loss by transpiration, and, if water is limiting, may reduce water stress and increase photosynthesis and yield of a C_4 species. This closure explains why some field experiments with elevated CO_2 levels unexpectedly show small yield increases for C_4 crops (Hatfield *et al.*, 2011).

CO$_2$ concentration in the atmosphere has been increasing steadily since the beginning of the Industrial Revolution, primarily because of increasing combustion of fossil fuels (derived originally from photosynthesis). The concentration in 1800 was roughly 280 ppm, but it increased to 414 ppm by 2019. Increasing atmospheric CO_2 levels have probably contributed to historical yield increases of C_3 species including soybean, wheat and rice, but future contributions from rising

CO_2 levels may be limited by deteriorating environmental conditions (high temperatures, lack of water) resulting from climate change.

Temperature

All of the metabolic processes that collectively make up plant growth, including photosynthesis, are affected by temperature. The response of these metabolic processes to temperature can be characterized by the classic minimum–optimum–maximum response partially illustrated for photosynthesis in Fig. 2.4. Photosynthesis is zero at a minimum temperature (below the lowest temperature shown in Fig. 2.4), it increases to a maximum rate at the optimum temperature (actually a range in temperature) and then it declines again to zero at the maximum temperature (above the range in Fig. 2.4). The cardinal points for photosynthesis are obviously not the same for C_3 and C_4 species, and there is some variation among species within the C_3/C_4 classification. Species with C_3 photosynthesis have an advantage over C_4 species at temperatures below roughly 15°C (59°F), while C_4 species do not reach their maximum rate (optimum) until temperatures reach approximately 30°C (86°F), giving them an advantage over C_3 species at temperatures greater than approximately 15°C (59°F) (Fig. 2.4). Photosynthesis is the driving force behind the production of yield, but the temperature responses shown in Fig. 2.4 do not necessarily predict yield in the field. This failure occurs because the temperature response of other important plant growth processes that also influence yield may differ from the response of photosynthesis. Temperatures that produce the highest rate of photosynthesis may not produce the highest yield.

One important example of differential temperature responses involves the effect of temperature on the rate of plant development – the time it takes the

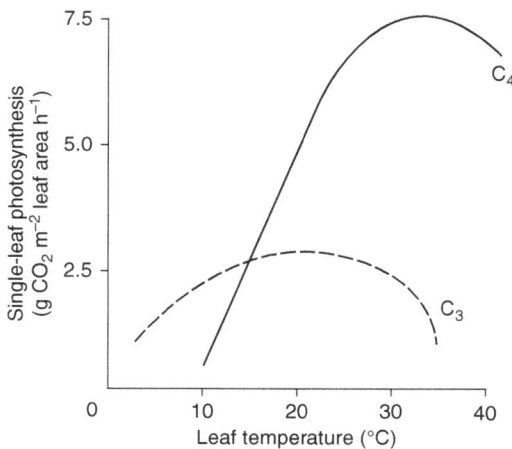

Fig. 2.4. Response of single-leaf photosynthesis of a C_3 and C_4 species to variation in leaf temperature at ambient CO_2 concentrations. Note: 10°C = 50°F and 30°C = 86°F. (From Conner *et al.*, 2011, p. 268.)

plant to progress through its life cycle from seedling emergence to maturity. The length of the seed-filling period is of particular interest because it is directly related to yield (we will discuss this relationship at some length in Chapter 3). Lower temperatures slow the rate of development, lengthening the effective filling period (an estimate of the length of the seed-filling period) (Fig. 2.5) and allowing more time for the accumulation of yield. Consequently, temperatures below the optimum for photosynthesis (Fig. 2.4) may increase yield by allowing more time for the seeds to grow. Temperatures in the optimum range for photosynthesis result in higher rates of photosynthesis, but a shorter seed-filling period, which could result in lower yield.

Temperature extremes also affect yield. Exposing plants to excessively high temperatures during flowering and seed set can disrupt pollination or fertilization, thereby reducing the number of seeds produced and very likely reducing yield. These reductions in yield, which can be catastrophic, are not related to the rate of photosynthesis.

We have discussed two scenarios where temperature influences yield in ways that are not related to the photosynthesis response curves in Fig. 2.5. These curves describe the effect on photosynthesis, but they cannot be used to reliably predict temperature effects on yield; we will discuss these interactions in detail later in this chapter and in Chapter 3.

In summary, the rate of photosynthesis in a farmer's field is determined by species (C_3/C_4), possibly by variety within a species; environmental conditions, principally the levels of solar radiation, CO_2, temperature and the availability of water; and the proportion of the incident solar radiation intercepted by the crop community (see Chapter 3). The availability of mineral nutrients also affects photosynthesis (N deficiency is perhaps the most obvious), but mineral nutrition is beyond the scope of this book.

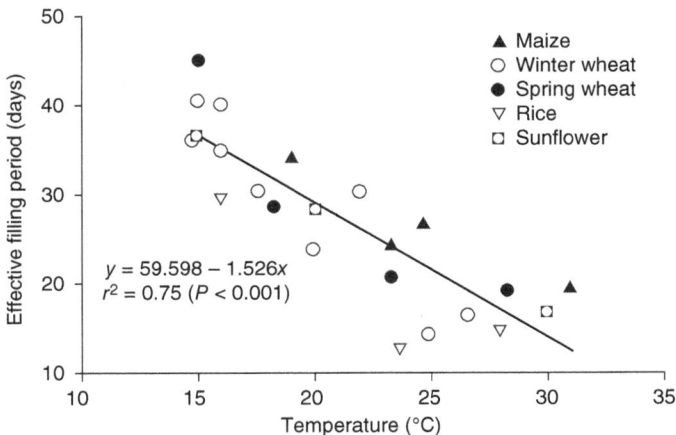

Fig. 2.5. The relationship between temperature and seed-fill duration for several grain crop species. Note: 10°C = 50°F and 30°C = 86°F. (From Egli, 2004.)

The environment that maximizes photosynthesis (and often yield) is one where temperature is in the optimum range for photosynthesis and adequate water supplies and high levels of solar radiation are available. CO_2 levels above ambient would be a bonus for C_3 crops but CO_2 concentration does not vary much by location or on a day-to-day or a year-to-year basis and we cannot manipulate it in the short run; we can only accept what the atmosphere provides.

The fact that yield is produced by photosynthesis implies that environmental conditions, genetic improvement and management practices that affect yield must express their effect on the rate and/or the duration of photosynthesis. There are some exceptions to this statement, but, by and large, it is true; the crop manager is managing photosynthesis.

Respiration

Respiration is the metabolic process that breaks down the products of photosynthesis to make energy available, in the form of high-energy compounds (ADP/ATP, etc.), to drive all the metabolic processes necessary for plant growth. Respiration (Eqn 2.2) consumes simple carbohydrates and O_2 and releases CO_2 to the atmosphere, which is the exact opposite of photosynthesis (consumes CO_2 and releases O_2) (compare Eqn 2.2 with Eqn 1.1):

$$C_6H_{12}O_6 + 6O_2 \rightarrow 6CO_2 + 6H_2O + \text{Energy} \qquad (2.2)$$

The synthesis of all the compounds that make up a plant (amino acids, protein, starch, lignin, cellulose, complex sugars of various forms, DNA, RNA, fatty acids, oil, etc.) from carbon skeletons requires energy and respiration provides this energy. The uptake of nutrients (NO_3^-, P, K, etc.) from the soil solution and the reduction of NO_3^- to NH_4^+ in the leaves or the fixation of N_2 from the atmosphere by rhizobia in the nodules of legumes also require energy from respiration.

It is important to realize that CO_2 evolved by respiration is not wasted; respiration is a vital, necessary part of plant growth. Historically, the apparent opposite nature of respiration and photosynthesis led to the incorrect idea that respiration was 'bad' because the evolution of CO_2 must reduce growth and yield compared with the 'goodness' of photosynthesis where the uptake of CO_2 fuels plant growth. This viewpoint is completely wrong – respiration is just as essential for plant growth as photosynthesis. It is important to note that the photorespiration associated with photosynthesis in C_3 species is not related to the respiration discussed in this section. Photorespiration is 'bad', but the respiration discussed in this section is 'good' – it is essential for plant growth.

We do not know as much about the variation in respiration in the field as we know about photosynthesis because it is harder to measure than photosynthesis. Some scientists measure CO_2 evolution from the plant in the dark and then estimate respiration during the day by adjusting the dark evolution of CO_2 to daytime temperatures. The few data available suggest that respiration may account for 40 to 50% of the carbon taken up by photosynthesis of the crop over its life cycle (Conner *et al.*, 2011, p. 302).

Respiration has two functional components: (i) respiration responsible for growth of the plant (the accumulation of dry matter by leaves, stems, roots and seeds) (growth respiration); and (ii) respiration that simply maintains the existing plant tissues (maintenance respiration). The energy from growth respiration is responsible for the synthesis of new tissues that makes up the plant, the accumulation of minerals and the acquisition of N; in short, all of the metabolic processes associated with plant growth. Consequently, the rate of growth respiration is directly related to the rate of photosynthesis and the rate of growth, but the respiration required to synthesize a gram of plant tissue is not affected by temperature. Plants that are growing rapidly will have high rates of growth respiration. Growth cannot occur without growth respiration, so the loss of CO_2 from growth respiration is inevitable and an essential part of the growth process.

The respiration needed to synthesize a unit weight of plant tissue, i.e. the energy cost, depends upon the composition of the tissue. Penning de Vries *et al.* (1974) carefully evaluated the energy cost of synthesizing various plant components from glucose and found that the synthesis of starch required the least energy (characterized by the grams of glucose required to synthesize one gram of product, 1.21 g glucose g starch^{-1}), oil required the most glucose (2.71 g g^{-1}) and protein was intermediate (2.48 g g^{-1}). More respiration and CO_2 evolution are required to synthesize oil with its higher energy content than starch with its lower energy content. More glucose would be required to produce one gram of soybean seed (containing high oil and protein concentrations) than one gram of wheat seed (high starch concentration) (Table 1.1). Differences in plant composition have a direct effect on plant growth and yield because the amount of plant tissue that can be produced from a given level of photosynthesis depends upon the composition of the tissue. For example, over 86% more glucose (i.e. more respiration) is required to synthesize one kilogram of soybean seed (2.64 g glucose g seed^{-1}) (assuming the soybean obtains all of its N from N_2 fixation) than one kilogram of maize seed (1.42 g glucose g seed^{-1}) (Conner *et al.*, 2011, p. 300). If the two crops had similar plant growth characteristics and equal photosynthesis, maize yield would be 86% higher than soybean yield.

The energy cost of synthesizing seed tissues is not the only reason why there are species differences in yield, but it is important. The development of higher-yielding grain crop varieties is often associated with decreases in seed protein concentration. Replacing protein (relatively high energy requirement) with compounds with lower energy requirements (e.g. starch) allows the synthesis of more seed tissue from the same amount of photosynthate, thereby increasing yield. Interestingly, soybean yield and protein per unit area could probably be increased by selecting aggressively for low oil concentration, but this would, of course, reduce the crop's value as a source of oil for human consumption.

The plant must expend energy to acquire N from its environment and the energy cost depends upon the source of the N. Using NO_3^- from the soil solution requires less energy than acquiring it via N_2 fixation in the nodules of legumes (see Conner *et al.*, 2011, pp. 298–300 for a detailed explanation). The higher energy cost of N from N_2 fixation suggests that the yield of legumes could be increased by as much as 30% by replacing N from N_2 fixation with N from NO_3^- in the soil solution. These theoretical calculations explain why legumes will always use NO_3^- from the soil solution before N from N_2 fixation; the 'cheaper' source of N will provide more growth. The same logic suggests that applying N fertilizer to soybean (thereby eliminating N_2 fixation) will increase yield as shown in some field experiments (Cafaro La Menza *et al.*, 2019). Other field experiments, however, did not show any yield response to N fertilization. Perhaps other aspects of the field environment were limiting yield in those experiments. Whether the yield increase compensates for the cost of the N fertilizer and possible environmental degradation are questions a producer must answer when contemplating this avenue to increase yield.

Maintenance respiration produces the energy required to maintain existing plant tissues. Energy produced by maintenance respiration drives protein turnover (breakdown and resynthesis of proteins and enzymes), maintains ion gradients across membranes and fuels intracellular transport processes, among other maintenance activities. Maintenance respiration is an essential component of plant growth, just like growth respiration, but it usually comprises a much smaller proportion of total respiration. It is important to note that the division of respiration into growth and maintenance respiration is purely a functional division related to the use of the energy produced; it has nothing to do with the biochemical process of respiration.

The amount of maintenance respiration is directly related to the weight of the plant (the amount of tissue that needs to be maintained), whereas growth respiration is related to the rate of growth (production of new plant tissue) and not to the tissue already in place. Relating maintenance respiration to the weight of the plant suggests that increasing vegetative plant size without a corresponding increase in canopy photosynthesis may reduce plant growth and yield. This logic predicts that an early-maturing variety with a smaller vegetative plant mass may have a yield advantage over a later-maturing variety with a larger vegetative mass, assuming no difference in community photosynthesis. This does not happen, as we shall see in Chapter 3, probably because maintenance respiration rates of older tissues lower in the plant canopy are reduced (see Hay and Porter, 2006, p. 122). The rate of maintenance respiration increases as temperature increases, providing one possible explanation for the widely held belief that high night temperatures increase respiration and reduce yield. There may be, however, other mechanisms responsible for this belief, such as high temperatures shorten the seed-filling period, an issue that will be discussed in Chapter 3.

In summary, respiration is responsible for the loss of CO_2 from the plant, which is a reversal of the fixation of CO_2 by photosynthesis. Respiration, however, is every bit as essential for plant growth as photosynthesis – the production of simple sugars by photosynthesis is only the beginning of plant growth. Converting those simple sugars into the compounds making up plant tissue, maintaining the integrity of those tissues and acquiring N and other mineral nutrients are driven by energy from respiration. Respiration is an essential part of plant growth.

Leaf Senescence

Senescence, defined as 'the series of events concerned with the cellular disassembly in the leaf and mobilization of materials released during the process' (Thomas and Stoddart, 1980), represents the final stage in the life of a leaf. Chlorophyll, Rubisco and other leaf proteins are broken down and translocated out of the leaf during senescence, leading to a decline in photosynthesis and eventually the characteristic leaf yellowing associated with approaching maturity. Leaves at the bottom of the plant may senescence during the vegetative growth phase, before the beginning of seed filling, but senescence of the upper leaves and the decline in community photosynthesis (canopy apparent photosynthesis) do not usually begin until vegetative growth has stopped (no more new leaf production) and the seeds are growing (Fig. 2.6). Surprisingly, the decline in photosynthesis often starts remarkably early in the seed-filling period. Senescence is usually complete (and leaves are abscised in legumes) when the seeds reach their maximum weight at physiological maturity, but there are exceptions to this pattern and green leaves may remain on the plant at maturity. The 'new' higher-yielding soybean varieties in Fig. 2.6 (open circles, ○) always had higher photosynthesis during seed filling than the older, lower-yielding varieties (closed squares, ■), which probably contributed to their higher yields.

Senescence presents an interesting enigma – when the seeds start growing and the crop is finally producing yield, the plant starts to disassemble its photosynthetic apparatus, destroying its capacity to produce the assimilate needed for seed growth. Photosynthesis is the source of yield, but senescence gradually reduces photosynthesis just when it seems to be needed the most (Fig. 2.6).

The effects of declining photosynthesis during seed filling are somewhat negated by the remobilization of N and other breakdown products from the leaves to the seeds. This remobilized N accounted for 20% (early-maturing variety) to 100% (late-maturing variety) of the seed N at maturity (Zeiher *et al.*, 1982), thereby reducing the need to expend energy during seed filling to accumulate N from the soil solution or by N_2 fixation. Senescence and the associated remobilization also reduce the N and other nutrients in the crop residues left after harvest. The peak N concentration in soybean leaves, for example, was roughly $4 \text{ g} 100 \text{ g}^{-1}$ (4%), which was reduced to around $2 \text{ g} 100 \text{ g}^{-1}$ (2%) when the leaves abscised (Zeiher *et al.*, 1982).

Senescence follows this same general pattern in all grain crops. There is little evidence that senescence is triggered by the 'demand' of the seed for N;

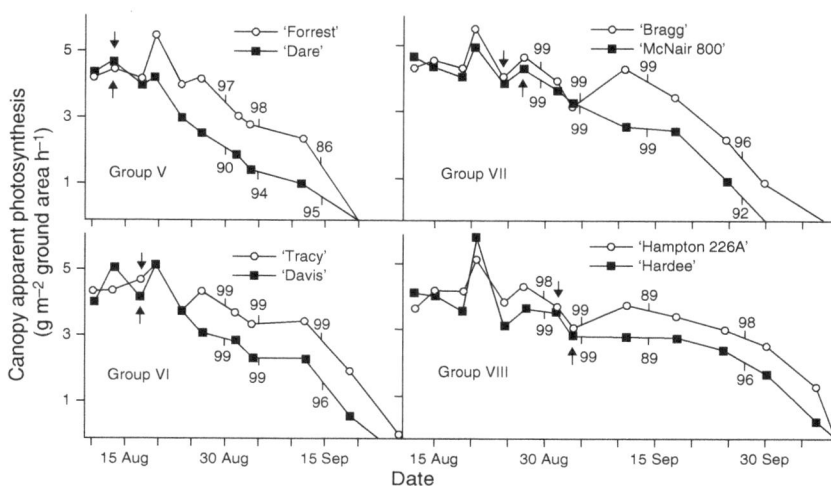

Fig. 2.6. The decline in community photosynthesis associated with leaf senescence of soybean varieties from Maturity Group V through VIII. The arrows indicate the beginning of seed filling. The two varieties in each maturity group represent a recent release (open circles, ○) and an older variety (closed squares, ■). Numbers beside the lines indicate the percentage solar radiation interception. (From Wells *et al.*, 1982. Used with permission from John Wiley and Sons.)

consequently, it is not restricted to crops that produce high-protein seeds (soybean, groundnut and other grain legumes). Senescence occurs in a similar manner in maize and other cereals producing low-protein seeds. If, for any reason, senescence is not complete when the seeds are mature, the presence of green vegetative tissues may cause harvest problems.

Senescence is influenced by environmental conditions during seed filling. Since seeds cannot grow without assimilate from photosynthesis, variation in the pattern of senescence can affect yield. The rate of senescence is accelerated by high temperature and N or water stress during seed filling. There is some evidence in soybean that only short periods (perhaps as few as 3 days) of water stress will trigger an acceleration of senescence that cannot be reversed by eliminating the stress (Brevedan and Egli, 2003). Leaf diseases also accelerate senescence, which means that control of leaf disease may delay senescence, prolong seed filling and increase yield. Since the seeds depend on the mother plant for the raw materials for seed growth, the length of the seed-filling period and the rate of leaf senescence are intimately linked. Consequently, if environmental stress affects senescence and shortens the seed-filling period, yield is usually reduced. We will discuss these effects in more detail in Chapter 3.

The senescence process is also under genetic control as shown by variety differences in the rate of senescence that are consistent across years. Plant breeders have also shown that variation in senescence is a heritable trait, but direct selection for this trait has proven difficult. Interestingly, selection for

high yield by plant breeders resulted in a delay in senescence and a longer
seed-filling period in several crops as illustrated by comparisons of old and new
soybean varieties in Fig. 2.6.

In summary, senescence is the final phase of the crop growth cycle; when it
is complete the production of yield is finished. Senescence efficiently recycles
N and other nutrients from the leaves and other vegetative plant parts to the
seeds. Environmental effects on senescence can prematurely terminate seed
filling, reducing yield. Clearly, yield is not immune to environmental insult
until the seeds reach their maximum weight at physiological maturity.

Seed Growth

We must consider the capacity of seeds to grow and accumulate the oil, protein and complex carbohydrates that give seeds their value to fully understand
the production of yield. After all, the yield of grain crops is the seeds harvested
at maturity. Previously we discussed photosynthesis, the fundamental process
that produces plant growth and provides the raw materials the seeds need to
grow. Production of raw materials is only part of the story; utilization of the
raw materials by the seeds is equally important and cannot be neglected. Seeds
are not simply empty containers to be filled by the mother plant; instead, they
are metabolic powerhouses that, to a remarkably large extent, control their
own growth and development. Control of seed growth by the characteristics
of the seed gives the seed an important role in the yield production process. We
cannot completely understand how crops produce yield without considering
the seed and seed growth processes.

Seeds of grain crops vary widely in their shape, colour, size, composition
and morphology. The harvested seed of the grasses (e.g. maize, rice, wheat) is a
caryopsis (a single-seeded fruit with the pericarp fused to the testa (seedcoat));
the legumes produce a non-endospermic seed (there is no endosperm tissue
present in the mature seed) while sunflower produces an achene (an indehis-
cent fruit containing a single seed that nearly fills the pericarp). For simplicity,
we will use the general term 'seed' for all species.

The major grain crop species exhibit a 20-fold variation in seed size (i.e.
weight per seed), with average sizes ranging from 563 mg seed^{-1} for groundnut
(peanut) to 28 mg seed^{-1} for rice and sorghum (Table 2.2). Maize seeds (302 mg seed^{-1})
are typically larger than soybean seeds (202 mg seed^{-1}), while wheat and barley
seeds are smaller (40 mg seed^{-1}). There are some species that produce huge seeds
(broad bean produces a seed of more than 1000 mg) and some very tiny seeds (a
flax seed weighs only 8 mg). There is also variation in seed size within a species
represented by the range in Table 2.2. The production environment influences
seed size, so the seed size of a single variety varies among locations and years.
Seeds on the same plant are not all the same size as shown by a nearly twofold
variation in size among seeds from a single soybean plant (Egli *et al.*, 1987).

Table 2.2. Typical seed sizes (weight per seed) of important grain crops. (Adapted from Egli, 2017, p. 46.)

Species	Number of varieties	Mature size[a]	
		Mean (mg seed^{-1})	Range (mg seed^{-1})
Grasses			
Wheat	26	41	23–55
Barley	13	38	22–55
Rice	12	28	20–50
Sorghum	9	28	19–37
Maize (inbreds)	22	228	86–322
Maize (hybrids)	10	302	229–410
Legumes			
Soybean	21	202	84–484
Bean	20	345	190–545
Cowpea	3	73	32–122
Groundnut (peanut)	2	563	500–625
Oilseeds			
Flax	2	8	7–8
Sunflower	7	54	39–75

[a]Seeds lb^{-1} = 453.6/[(mg seed^{-1})(0.001)]. Seeds kg^{-1} = (Seeds ib^{-1}) (2.205)

The variation in seed size among species and some of the variation among varieties within a species is under genetic control. Size is not closely associated with plant or seed characteristics or seed composition; large- and small-seeded species can be found in both grasses and legumes (Table 2.2). However, as a group, legumes tend to have larger seeds than grasses. Perhaps surprisingly, as we shall see later in Chapter 3, genetic differences in seed size (within and among species) are not related to yield. The variation in seed size created by the production environment during seed development (the environmental component of seed size) may be related to yield.

In spite of all the variation in seed characteristics, the patterns of seed growth and development are uniform among species, so we can develop descriptions of the role of the seed in the production of yield that apply to all species. We will not have to discuss the growth characteristics of individual species, which will make our investigations in the next section much simpler.

Growth of individual seeds

The curve representing the accumulation of dry weight by a soybean seed was produced by harvesting pods developing from flowers that opened on the same day and measuring seed fresh and dry weight (Fig. 2.7). This curve does not represent an average of all the seeds on the plant; it only represents the growth

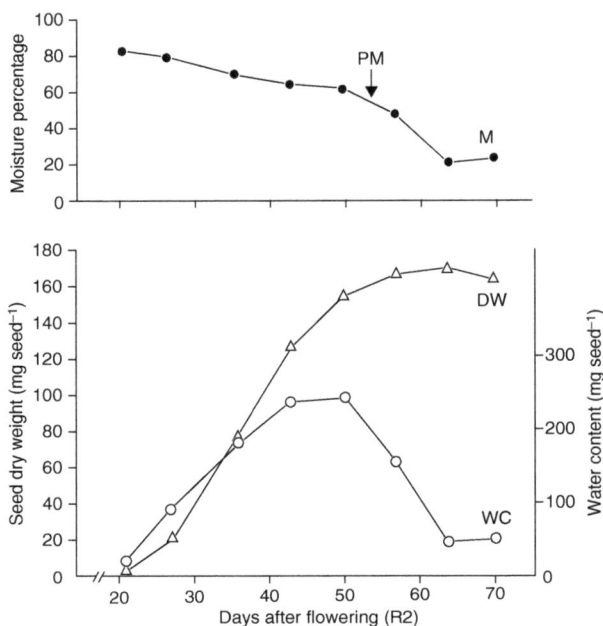

Fig. 2.7. Seed dry weight (DW), water content per seed (WC) and water concentration (M) of an individual soybean seed developing in a field environment. PM, physiological maturity. (From Fraser *et al.*, 1982. Used with permission from John Wiley and Sons.)

of a seed developing from flowers of the same age. It is easier to understand seed growth characteristics when following the growth of an individual seed instead of all the seeds on the plant, because they develop at different times.

Dry weight accumulates slowly during the initial lag phase, which is followed by a period when the growth rate (mg seed^{-1} day^{-1}) is constant (linear phase of growth) before the rate slows and stops at physiological maturity (maximum seed dry weight). Seeds of all species follow this pattern, although the rate of growth during the linear phase and, to a lesser extent, the length of the growth period vary among species and varieties within a species.

The initial lag phase from pollination to the beginning of the linear growth phase is a time of rapid cell division that produces all of the seed structures. Cell division stops at the beginning of the linear growth phase and the increase in seed size during the linear phase is a result of an increase in cell volume. The seed water content (mg H$_2$O seed^{-1}) increases during the linear growth phase, driving the increase in cell volume (Fig. 2.7). Cell volume (and seed volume) and seed water content reach a maximum before the seed stops accumulating dry matter at physiological maturity. The initial seed water concentration is high, roughly 80% (fresh weight basis), but it declines steadily during seed growth and reaches a concentration at physiological maturity that is characteristic

Table 2.3. Species variation in seed water concentration at physiological maturity. (From Egli, 2017, p. 25.)

	Caryopsis		Non-endospermic true seed
Species	Water concentration (g kg^{-1})[a]	Species	Water concentration (g kg^{-1})
Wheat	370–437	Soybean	550–600
Maize	337–377	Bean	520–535
Oat	450	Broad bean	510–600
Barley	420–480	Pea	550
Triticale	400	Chickpea	600
Pearl millet	350	White lupin	600–650
Sunflower[b]	380–410		
Sorghum	320		

[a]Water concentration in g kg^{-1} × 0.10 = water concentration in % (fresh weight basis).
[b]Seed is an achene.

of each species (Table 2.3). The seed water level declines rapidly after physiological maturity until it reaches a minimum level.

At physiological maturity the seed is no longer connected to the plant's vascular system, so seed dry weight and yield are at a maximum. The decline in moisture after physiological maturity depends upon environmental conditions since the seed is in storage on the plant. The moisture level declines rapidly if it is dry, but, if there is enough rain, the moisture concentration may actually increase and, in extreme situations, the seed may germinate on the plant. The rate of water loss after physiological maturity is faster in soybean and wheat than it is in maize (Egli, 2017, p. 20), probably because the husks around the ear are more restrictive of water movement than the structures surrounding wheat and soybean seeds.

Seed growth in crop communities

The growth of a single seed was the focus of our discussion so far, but of course the production of yield requires the growth of many seeds: a maize yield of 15,677 kg ha^{-1} (250 bu acre^{-1}) requires 58.1 million seeds ha^{-1} (23.5 million seeds acre^{-1}) assuming a typical weight per seed. The growth of each of these seeds follows the pattern just described for a single seed (Fig. 2.7), but they do not all start and stop growing at the same time. Interestingly, the starting time of individual seeds is more variable than the ending time. Pod initiation, and presumably the initiation of seed growth, occurred over a roughly 40-day period in an indeterminate soybean variety (roughly 30 days in a determinate variety) (Egli and Bruening, 2006), but most of the pods (~80%) were initiated in less than half of the 40-day period. The loss of green colour from the pod (the first visual indication of maturation), however, occurred over only an

11-day period (Spaeth and Sinclair, 1984). The last maize kernels on the tip of the ear started growing 4 to 10 days after the first kernels at the base of the ear (Tollenaar and Daynard, 1978; Bassetti and Westgate, 1993), a much shorter period than soybean. It seems likely that most other grain crop species would fall between these extremes.

Rate and duration of seed growth

The rate of growth during the linear growth phase ($mg\,seed^{-1}\,day^{-1}$) and the length of the seed-growth period (seed-fill duration in days) usually characterize seed growth. In other words, the final seed size (weight per seed) is a function of a rate expressed over a finite time. Seed growth rate (SGR) is estimated from the dry weight of repeated samplings during the linear phase of seed growth.

The slow growth during the initial and finial lag phases (Fig. 2.7) makes it difficult to accurately estimate the duration of seed growth, i.e. when the seed actually starts and stops growing. For that reason, the effective filling period, defined as the final seed size divided by the SGR, i.e. how long will it take the seed to reach its maximum size growing at the linear rate (Egli, 2017, p. 44), is often used as an estimate of the duration of seed growth (Fig. 2.5). Whole-plant growth stages are also used to estimate the duration of seed filling. Growth stage R5 to R7 in soybean or silking to physiological maturity in maize describe the duration of seed filling on a whole-plant basis. Other estimates of the duration of seed fill are discussed in Egli (2017, pp. 43–45).

Seeds rely on the mother plant for raw materials for growth, principally sucrose and a few amino acids; consequently, the supply of these raw materials could influence the SGR. Although the supply is important, the SGR is remarkably consistent across environments because the supply to the individual seed is buffered by the adjustment of seed number to the supply of sucrose. This adjustment maintains a relatively constant supply of sucrose per seed in most conditions unless there is a large change in supply when seed number can no longer adjust. Consequently, SGR is not affected as much by variation in environmental conditions as other plant growth characteristics. Interestingly, SGR is more sensitive to the supply of sucrose than it is to N (amino acids); in fact, soybean seeds grow very well in culture systems with very low levels of N, but they do not accumulate much protein. We will discuss these relationships in detail in Chapter 3.

Seed growth rate is sensitive to temperature, decreasing steadily as the temperature drops below 22°C (72°F) (Egli, 2017, p. 55). Excessively high temperatures would also reduce SGR, but less is known about the response to high temperatures. Interestingly, SGR is not very sensitive to water stress (Egli, 2017, pp. 53–54). Water stress could affect the metabolic ability of the seed to grow, or it could affect the supply of raw materials from the mother plant, i.e. reductions in photosynthesis. Research suggests that moderate water stress does not affect the metabolic capacity of the seed to grow. Severe stress would, of course, reduce SGR.

Variety or species differences in SGR that are consistent across years and/or locations indicate that SGR is under genetic control. The genetic differences in seed size discussed previously are usually a result of differences in SGR (Fig. 2.8). Big seeds grow rapidly (high SGRs) while little seeds grow slowly (low SGRs). Restrictions on the length of the crop growth cycle (often due to low temperatures) necessitate this relationship. For example, a plant producing a 300 mg seed (typical of maize) with an SGR of 1.4 mg seed^{-1} day^{-1} (typical of wheat) would require a 214-day effective filling period, obviously an untenable combination. The seed size–SGR relationship results in a relatively uniform length of the seed-filling period regardless of seed size. The close association between seed size and SGR explains why genetic differences in seed size are rarely related to yield. We will discuss this at length in Chapter 3.

Yield is always a function of a rate of growth expressed over a certain time interval, so the length of the seed-filling period is an important seed growth characteristic. Seeds that grow for a long time have a greater opportunity to accumulate dry matter – resulting in higher yield – than seeds that grow for a shorter period.

Since yield is often related to the duration of seed filling, answering the question 'why does the seed stop growing?' helps us understand the seed-fill duration–yield relationship. There are two answers to this question. First, seeds can grow

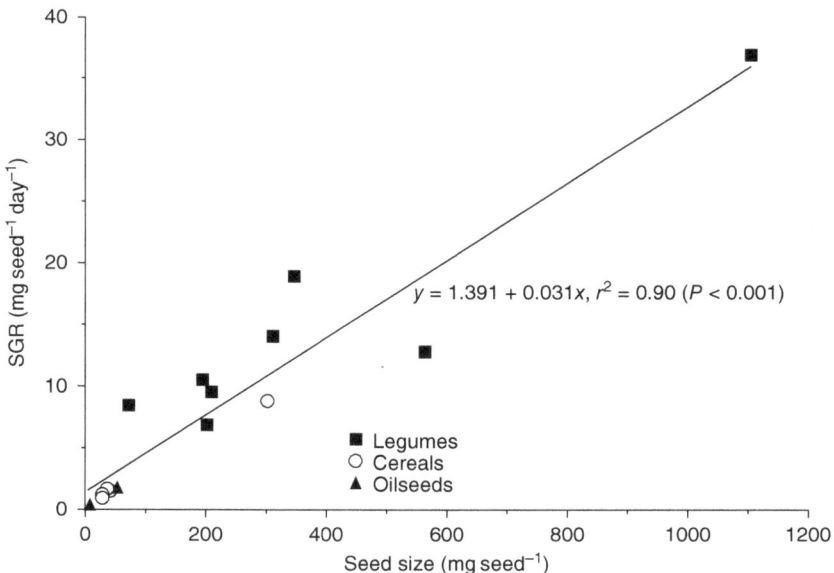

The equation shown on the graph:
$$y = 1.391 + 0.031x, \quad r^2 = 0.90 \ (P < 0.001)$$

Legend:
- ■ Legumes
- ○ Cereals
- ▲ Oilseeds

Fig. 2.8. The relationship between seed size (mg seed^{-1}) and seed growth rate (SGR) (mg seed^{-1} day^{-1}) for five cereal species (wheat, barley, rice, sorghum and maize), seven legume species (soybean, bean, pea, field pea, broad bean, cowpea and groundnut (peanut)), and two oilseed species (flax and sunflower). (Data from Egli, 2017, p. 46.)

only when they have a continuing supply of raw materials from the mother plant, so the completion of leaf senescence signals the end of seed growth. The second answer relates to the ability of the seed to continue to increase in volume or size. If reproductive structures (pod walls or carpels, glumes, spacing of the seed on the cob of maize) physically limit the increase in volume, the seed will stop growing even though green leaves are still supplying raw materials for seed growth (see Egli, 2017, pp. 70–75 for a more detailed discussion).

Seed-fill duration is sensitive to temperature, increasing by roughly 1.5 days $°C^{-1}$ (0.8 days $°F^{-1}$) as temperature decreases below roughly 30°C (86°F) (Fig. 2.5) This temperature response explains why high yields of many crops occur in relatively cool climates. The crop has more time to accumulate yield when temperatures are lower.

Since seed growth depends upon the supply of raw materials from the mother plant, acceleration of leaf senescence by, for example, leaf disease, nutrient deficiency or water stress will lead to a premature cessation of seed growth, a shorter seed-filling period and lower yield. As mentioned previously, water stress during seed filling is a frequent cause of accelerated leaf senescence, shorter seed-filling period and reduced yield.

There are consistent differences in seed-fill duration among varieties within a species in most grain crops, an indication of genetic control (Egli, 2017, pp. 63–64). Plant breeders were able to increase seed-fill duration via direct selection. Conversely, selecting for higher yield often resulted in longer seed-filling periods. Some higher-yielding varieties owe their high yield to longer seed-filling periods. The extra work required to measure seed-fill duration and the large environmental effects on it have so far stymied efforts to increase yield by direct selection for longer seed-filling periods.

Physiological maturity

Physiological maturity is defined as when the seed reaches its maximum dry weight, making it an important growth stage because no yield is produced after physiological maturity. The production of yield by the crop is finished at physiological maturity. Environmental stress, disease or insect infestations after physiological maturity will not affect yield per se, but they can reduce the harvested yield if they increase harvest losses, or they may reduce grain quality. The weight of the seeds at physiological maturity is the yield that the crop produced, and it is the focus of crop physiologists when they study the processes that determine yield, but it is harvested yield that puts money in the producer's pocket.

The SGR slows down as the seed approaches physiological maturity (see Fig. 2.7), so not much yield is accumulated in the last few days before physiological maturity. Consequently, the estimated date of physiological maturity can vary by several days with only minimal effects on yield. Although the end of seed growth is less variable than the beginning, all of the seeds on the plant do not reach physiological maturity at the same time. Consequently, the occurrence of physiological maturity is usually based on a proportion of the seeds reaching this growth stage.

There may also be variation in the timing of physiological maturity within a field; plants on hilltops that experience drought stress because of the shallower soils will probably reach physiological maturity before plants in other areas that have access to more moisture. This variation complicates scheduling management practices at physiological maturity.

Accurate, simple and easy-to-use indicators of physiological maturity are useful management tools. Measuring seed dry weight to determine when it reaches a maximum would, theoretically, provide an indication of physiological maturity, but the variation in weight from sample to sample requires complicated statistical processing to identify the date of physiological maturity. Researchers use seed dry weight, but it is not a practical tool that can be used in producers' fields. Seed moisture levels, as mentioned earlier, provide a good indication of physiological maturity, but measuring seed moisture is complicated and often requires several days. Repeated samples are necessary to determine when the seed reaches the moisture level associated with physiological maturity, so this method is not widely used by producers. Methods based on colour changes of the seed or reproductive structures provide immediate, relatively accurate estimates of physiological maturity and they are widely used on a field scale. Seed characteristics are better indicators of physiological maturity than other plant parts, such as leaf yellowing or leaf abscission. Leaf characteristics are often associated with the end of seed growth (seeds cannot grow without a source of raw materials), but seeds may mature when the leaves are green and assimilate is still available; consequently, leaf characters are not useful as indicators of physiological maturity.

Soybean

A soybean seed reaches physiological maturity when it first turns yellow. Growth stage R7 (one pod on the main stem has reached its mature pod colour) (Fehr and Caviness, 1977) is an acceptable whole-plant measure of physiological maturity in soybean. Only 26% of the seeds were completely yellow at R7, but 70% showed some degree of yellowing and were probably within a few days of reaching their maximum weight. Consequently, yield determined from harvests of several field communities at growth stage R7 did not differ significantly from yield estimated at full maturity (growth stage R8, 95% of pods mature) (TeKrony *et al.*, 1981). Growth stage R7 is an unambiguous, easily observed growth stage and can be used in producers' fields to give an immediate indication of physiological maturity. The date of physiological maturity is usually taken as the day when 50% or more of the plants in a ten-plant sample (consecutive plants in the row that have a normal main stem) are at R7 (Fehr and Caviness, 1977).

Maize

Two seed characteristics, the black layer and the milk line, are used as indicators of physiological maturity in maize. The appearance of a black layer at the base of the seed signifies blockage of movement of sugars and amino acids into the seed and the end of seed growth (Daynard and Duncan, 1969). Hunter *et al.* (1991) found that maximum seed dry weight occurred when the black layer consisted of a thin, dark-brown band (usually less than 1 mm thick) reaching across the entire base of the seed (growth stage R6, Ritchie *et al.*, 1993).

The milk line is the line on the abgerminal face of the seed dividing the liquid and solid endosperm. There is no milk line on immature seed when all of the endosperm is liquid, but as the endosperm solidifies, the milk line moves down from the top of the seed until all of the endosperm is solid at maturity and again there is no milk line. Physiological maturity occurs when the milk line is near the seed's base (75% of the seed's length contains solidified endosperm) (Hunter *et al.*, 1991).

Other crops

Seed characteristics associated with physiological maturity have been identified for many other grain crop species. Some are based on the appearance of a dark closing layer at the base of the seed (sorghum, pearl millet) or when seed or reproductive plant parts reach their mature colour (sunflower, oat, wheat and barley). These relationships are discussed in more detail in Egli (2017, pp. 38–41).

In summary, seeds produce yield from the raw materials (sucrose and amino acids) supplied by the mother plant. Photosynthesis in the leaves is the source of these raw materials, but yield does not exist until the seeds use these raw materials to synthesize the oil, protein and complex carbohydrates that give seeds their value. Understanding the source of the raw materials is not enough; we must also consider the sink – the seeds. Understanding how seeds grow and respond to the environment provides key insights into how and why crop yield responds to management.

Water

The availability of water is the main yield-limiting factor in many environments where grain crops are grown. As discussed previously in this chapter, photosynthesis and water loss from the crop are intimately linked; photosynthesis cannot occur without water loss from the leaves. Crop plants cannot produce high yields without adequate supplies of water. Water is not just essential for plant growth and the production of yield, it is required in large quantities. Water use by a well-watered maize or soybean field in the US maize belt could be as much as 6.35 mm day^{-1} (0.25 in day^{-1}) ($63,500 \text{ l ha}^{-1} \text{day}^{-1}$ or $6791 \text{ gal acre}^{-1} \text{day}^{-1}$) on a warm, sunny July day with wind blowing when the maize or soybean leaves completely cover the soil surface. The use of 6.35 mm day^{-1}

(0.25 in day^{-1}) translates into 44.45 mm (1.75 in) per week, which can total as much as 508 to 762 mm (20 to 30 in) for the crop's life cycle, depending on its length. These large quantities help us understand why irrigated agriculture uses roughly 69% of the total global water withdrawals from streams and groundwater (Conner *et al.*, 2011, p. 384).

Rainfall, irrigation (if available) and stored soil moisture must be adequate to meet this demand to avoid stress and possible yield loss. Seldom does nature cooperate by providing 44.45 mm (1.75 in) every week, so the great challenge facing rain-fed agriculture is to match the constant, relentless daily use of water by the crop with intermittent rainfall to avoid stress. The capacity of the soil to store water plays a major role in balancing demand with the supply. It is not surprising that most producers are obsessed with the weather report during the summer unless they bought some peace of mind by purchasing an irrigation system.

Evapotranspiration (ET)

Water use by a grain crop (or any other crop) has two components: (i) water that evaporates from the soil surface (evaporation); and (ii) water that is lost from the leaves of the plant (transpiration). Transpiration involves water evaporating in the sub-stomatal cavity inside the leaf and moving, as vapour, out of the leaf through the stomata (see Fig. 2.1) into the atmosphere. These two processes are combined into evapotranspiration (ET) (evaporation from the soil surface plus transpiration from plant leaves). We can safely combine these two processes because, first, they both represent a loss of water from the crop community that must be replaced, and second, the basic mechanisms responsible for the loss are the same – liquid water is converted to vapour (evaporation) and the water vapour is transported out of the plant community to the atmosphere. Combining them is convenient because it is difficult to accurately measure soil evaporation and transpiration separately.

The effect of the environment on ET can be best understood by considering the two processes responsible for the water loss – evaporation of water and transport of the water vapour away from the crop. Environmental conditions affect both of these processes.

Evaporation of water requires energy; it takes 245 MJ to change 1 kg of water from a liquid to vapour (585 cal g^{-1}) (latent heat of vaporization) at 20°C (68°F). This energy requirement applies to evaporation of water from the soil surface or inside the plant leaf where transpiration originates. Incidentally, the energy used to evaporate water is given off when the water vapour condenses back to a liquid, such as when dew forms on plants at the earth's surface or condensation high in the atmosphere forms clouds. The energy to evaporate water comes from solar radiation, so the level of solar radiation plays an important role in determining the rate of ET.

The relatively large energy requirement for these transformations gives water an outsized role in regulating air temperature. If water is limiting

(ET is reduced and less energy is used to evaporate water), the excess energy will heat the air, the leaf and the soil. Examples of the linkage between air temperature and water abound in nature. Air temperatures are often higher in dry climates or during droughts in humid climates because the lack of water limits the energy used for ET and the excess energy is available to heat the air. Development of artificial turf for athletic fields provided a classic example of this relationship. Plastic 'grass' does not transpire, so air temperatures over those surfaces soared. When transpiration is limited by the availability of water, leaf temperature is often higher than air temperature, so measuring the difference between leaf and air temperature with an infrared thermometer provides an indication of the degree of moisture stress. Water is not only required for plant growth, its presence moderates air temperature, preventing extremes that could reduce plant growth and yield.

The second important component of ET is the transport of water vapour from the crop community to the atmosphere. Much of this transport is by diffusion. The rate of diffusion depends upon the difference in water vapour content between the air in the sub-stomatal cavity inside the leaf (usually saturated with water vapour) or just above the soil surface and the air around the leaf or near the soil surface (the diffusion gradient). The ET rate will be larger when the water vapour content of the air (humidity) is low (large gradient) and the ET will decrease as the humidity increases. The pressure due to water vapour is a measure of the water vapour content of the air, so the vapour pressure deficit (vapour pressure at saturation minus the vapour pressure of the atmosphere) is often used as a measure of the diffusion gradient. ET will generally increase as the vapour pressure deficit increases (assuming, of course, that there is adequate water in the soil).

As ET continues in still air, the water vapour content of the air around the leaf or near the soil increases, which decreases the vapour pressure gradient and slows the rate of diffusion and ET. Wind blowing through the crop canopy will replace this high-moisture air with drier air from the atmosphere, thereby restoring the diffusion gradient (restoring the vapour pressure gradient) and increasing ET. Turbulence (air moving up and down) of wind blowing across the crop community will also transport high-moisture air away from the crop (turbulent transfer). The air moving up (moist air) will be replaced by drier air moving down, thereby maintaining the vapour pressure gradient and the rate of ET. Consequently, ET will usually be higher on windy days. A 'rough' surface (maize field) creates more turbulence when the wind blows across it than a smooth surface (closely mown golf green).

Air temperature also influences the ET rate. The saturation vapour pressure of water (the maximum amount of water vapour in the air) goes up rapidly as the temperature increases. Consequently, increasing air temperature increases the water vapour content of the air inside the leaf, thereby increasing the vapour pressure gradient from the leaf to the air and increasing diffusion and ET.

To summarize, the highest rates of ET will occur when there is adequate water in the soil, it is sunny (high energy supply), warm with low humidity (larger diffusion gradient or vapour pressure deficit) and the wind is blowing (enhances turbulent transfer, which maintains the diffusion gradient and the vapour pressure deficit). Consequently, ET of irrigated crops (water is not limited) will be much higher in a desert climate (clear skies, high solar radiation, high temperature, dry air, often windy) than in a humid climate (moister air, probably cloudy skies and lower solar radiation, cooler temperatures). The water use from an irrigated crop in a humid climate will be higher than normal during a drought for the same reasons (fewer clouds, more solar radiation, drier air, higher temperatures). The water required to produce the same yield under irrigation in a desert will be higher than in a humid climate. Making the desert bloom with irrigation is not necessarily a good idea from the viewpoint of water-use efficiency. Often the desert blooms only because government subsidies reduce the cost of water to the producer.

Other aspects of the crop that can influence ET include the colour of the leaves, which could affect the amount of solar radiation absorbed (not reflected) by the leaf; only absorbed solar radiation provides energy for evaporation. A plant community that has an uneven, 'rough' surface (variation in plant height) will create more turbulence in the air moving across the community than a community with a relatively 'smooth' surface (e.g. soybean). A community with a 'rough' surface (e.g. maize) will probably have a slightly higher ET rate than a community with a 'smooth' surface when all other conditions are equal. The effects of colour and 'roughness' on the rate of ET are not nearly as important as environmental conditions (solar radiation, temperature, vapour pressure deficit and wind speed). We must not forget that the availability of water in the soil trumps all other factors affecting ET.

Seasonal changes in environmental conditions and variation in the size of the plants (primarily leaf area) are also responsible for changes in ET during the crop's life cycle (Fig. 2.9). The crops in Fig. 2.9 were grown in a dry summer environment and they were irrigated, so the availability of water did not affect ET. The grass curve (leaves always completely covering the soil) primarily represents the effect of seasonal changes in the environment on ET. The ET is low in the spring when solar radiation and temperatures are lower, and it increases to a maximum in July when solar radiation and temperature are at the seasonal maximum before declining in the autumn. The ET of the other four crops is below the grass curve early in crop development because the plants are small and their leaves do not cover the soil completely. Transpiration is limited by the small leaf area and low soil evaporation when the soil surface is dry (a common situation), so ET is lower. The plant roots are extracting water for transpiration from deeper in the soil profile, so transpiration is not usually affected by the dry soil surface. ET of these four crops increased as the plants produced more leaves, increasing ground cover, and as the environment supported higher ET (as shown by the grass curve). When there are enough leaves to completely cover the soil surface, ET is completely dependent upon the environment (solar radiation, temperature, wind speed and water vapour content of the air) and the availability of water.

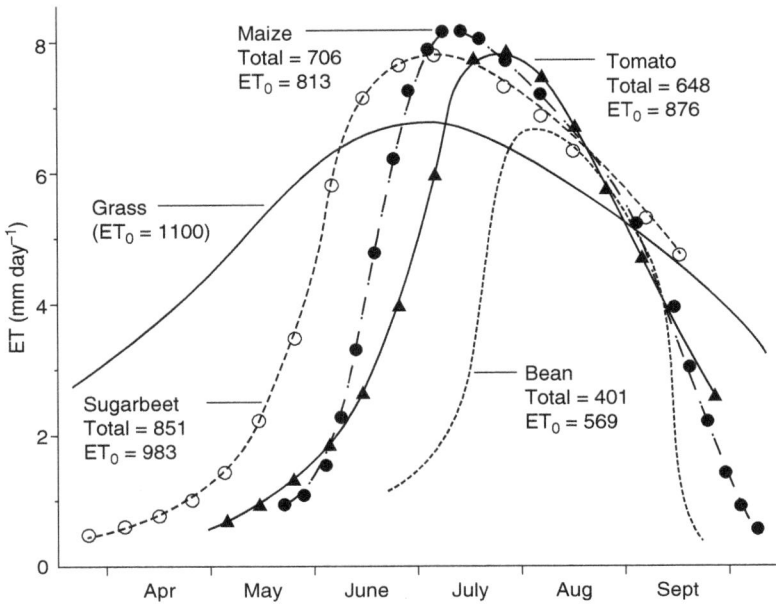

Fig. 2.9. Seasonal water use (evapotranspiration, ET) of five well-watered crop communities at Davis, California, USA, which has a dry-summer Mediterranean climate. Water use by grass represents the reference (potential) evapotranspiration (ET_0). The total ET and ET_0 during the complete life cycle are given for each crop. Note: 24.5 mm = 1.0 in, 8 mm = 0.31 in. (Adapted from Conner et al., 2011, p. 233.)

The maximum ET of three (maize, sugarbeet and tomato) of the four crops (excluding grass) (Fig. 2.9) was remarkably homogeneous, especially given the morphological differences among the three crops. The small differences could reflect variation in albedo, or canopy 'roughness' characteristics. The maximum rates of maize, sugarbeet and tomato were approximately 8 mm day^{-1} (0.31 in day^{-1}) reflecting the dry summer environment they were growing in. Bean had a lower maximum rate because it reached its maximum rate later in the season when the environmental conditions would not support a high rate as shown by the grass curve. The lower maximum ET of the grass is probably due to its 'smoother' surface. It is not surprising that the differences among species in total ET (grass, 1100 mm (43 in) to bean, 401 mm (16 in)) (Fig. 2.9) are mostly due to when and how long the crop grew, and not to their maximum rates.

The variation in seasonal water use of four irrigated grain crops grown in Idaho (Table 2.4) also illustrates the importance of the length of the growth cycle. Maize used the most water and had a longer life cycle than sorghum, wheat or soybean. There were substantial differences in water-use efficiency (g dry matter kg^{-1} water used) among the four crops, but they were not closely associated with the water used (total ET). Maize and sorghum had high water-use

Table 2.4. Water use and dry matter production of four well-watered grain crops grown at Kimberly, Idaho, USA. (Adapted from Gardner *et al.*, 1985, p. 95.)

Crop	Photosyn-thesis type	Growth period (days)	Total ET (mm)	Total ET (in)	Total dry matter (kg ha^{-1})	Water-use efficiency[a] (g kg^{-1})
Maize	C_4	135	658	25.9	17,000	2.59
Sorghum	C_4	110	583	23.0	14,000	2.49
Wheat	C_3	112	473	18.6	7,900	1.63
Soybean	C_3	113	599	23.6	8,300	1.42

[a]Grams of dry matter per kilogram of water used in evapotranspiration (ET).

efficiencies because they produced more dry matter (they both have C_4 photosynthesis), not because they used less water. Their water use was equal to or higher than that of soybean. A higher water-use efficiency does not necessarily mean that the crop used less water; in this example, they were more efficient because they produced more dry matter. These data collected under irrigation in a dry environment provide a dramatic illustration of the large amounts of water needed to produce grain crops (up to 658 mm or 25.9 in).

Reference or potential evapotranspiration (ET$_0$)

Reference evapotranspiration (ET$_0$) (originally referred to as potential evapotranspiration) was first defined in the late 1940s by C.W. Thornthwaite (1948) as 'water loss from an area that is completely covered by vegetation, without specifying the crop, from an area large enough so there are no oasis effects, when water is not limiting'. A more modern definition that incorporates the same concepts is 'evaporation from an extended surface of a short grass crop that fully shades the ground, exerts little or negligible resistance to the flow of water and is always well supplied with water' (Rosenberg *et al.*, 1983, p. 211). The key point of the concept of reference or potential evapotranspiration is that maximum water use by a crop, when water is not limiting, is controlled primarily by atmospheric conditions (solar radiation, wind speed, temperature and water vapour content of the air). In other words, atmospheric conditions primarily determine how much water a crop will use, unless ET is limited by a lack of water or by not having enough leaves to cover the ground. ET can be increased independently of ET$_0$ by oasis effects (increased ET from the horizontal movement of energy that occurs when a small crop area is surrounded by dry areas, i.e. similar to an oasis in a desert). Atmospheric conditions on a given day or at a specific location are more important in determining crop water use than the characteristics of the crop. The ET of a crop will approximate ET$_0$ unless water is limiting or the leaves do not provide complete ground cover. The oasis effect does not affect ET of large fields, but it can increase the ET of small research plots that are surrounded by dry areas.

The grass curve in Fig. 2.9 represents an estimate of ET_0. The seasonal variation is due to changes in environmental conditions, with the highest rate occurring in the summer when solar radiation and temperature are at a maximum. The maximum ET values of the four crops in Fig. 2.9 are somewhat greater than the maximum ET_0 values, which is a result of the effects of canopy characteristics on ET discussed previously. While there are small differences among species (see Fig. 2.9), the ET of a well-watered soybean crop with complete ground cover will essentially equal the ET of a maize crop under the same conditions. In practical terms, it will take more water to produce a crop of maize in an irrigated Arizona desert (high ET_0) than in central Illinois (lower ET_0), even though the yield may be the same. In terms of efficient water use, making the desert bloom is not always a good idea.

ET_0 is a very useful concept because it provides an estimate of maximum potential water use for a specific location or at locations in different climates, thereby defining the maximum amount of water needed to grow a crop without water stress. It is important to note that ET_0 is only the potential water use, not necessarily the actual water use by the crop. All sports fans are well acquainted with players with high 'potential' that is never realized. Their performance does not reach their potential, just as ET can be less than ET_0 because of incomplete ground cover by leaves or a lack of water in the soil.

Crop species can have a significant effect on ET when the conditions defining ET_0 are not met. For example, a species that reaches complete ground cover sooner will have a higher ET at that time than a species that lags behind. A species with deeper roots may have higher ET than a species with shallow roots when water is limiting. Management practices, principally row spacing and population, will affect when the crop reaches complete ground cover and ET during vegetative growth. We will discuss these relationships in greater detail in Chapter 3. In spite of this variation, we must not forget that maximum ET is mostly a function of environmental conditions, as represented by ET_0.

Measurement of evapotranspiration

Water use by crops (ET) is not easy to measure, especially on a daily basis, which has historically limited our knowledge and subsequently our ability to manage this important aspect of any crop production system. One method widely used is the soil water balance method (Eqn 2.3) which requires only measurement of precipitation and soil water levels to estimate ET:

$$ET = P - \Delta SWC - RO - DD \tag{2.3}$$

Precipitation (P) provides the input to the system, while losses of water from the system include surface runoff (RO) and deep drainage (DD), water that drains through the soil profile and becomes part of the groundwater. ET is represented then by precipitation minus the losses (RO and DD) adjusted for the change in soil water content (ΔSWC) over the measurement interval. In practice, RO and DD are often assumed to be zero (i.e. all the precipitation infiltrates and there is no deep drainage), not a bad assumption for an actively growing crop on relatively flat land in the summer in most climates where grain crops are grown. This technique requires repeated measurements of the soil water content throughout the soil profile, which recent developments in soil-water-measuring technology have made relatively easy. The inaccuracies in the measurement of soil water restrict estimates of ET to periods of several weeks or more. ET_0 could also be estimated with this technique if the conditions in the definition of ET_0 are met. Other methods to estimate ET provide more precise, short-term estimates but they require complex, expensive instruments (see Viney, 2005 for details).

ET_0 can also be estimated from the atmospheric characteristics that control water loss from the earth's surface. The most widely used methods to estimate ET_0 today are based on some form of the Monteith–Penman equation that includes all of the environmental variables that influence ET_0 (radiation, wind speed, temperature and water vapour content of the atmosphere). Simpler equations are available, but they are not theoretically correct. Techniques have been developed to estimate the variables in the Monteith–Penman equation from data available at standard weather stations (Ham, 2005).

Estimates of ET_0 approximate the maximum water use by a crop, so they provide a foundation for techniques to estimate irrigation timing, or they can be used to devise water-efficient cropping systems. Such techniques will be more important in the future, as water availability in agriculture, driven by climate change, becomes a larger issue.

In summary, grain crop productivity is, in large part, determined by how closely the supply of water matches the daily relentless use by the crop. The upper limit of water use is set by the ET_0, which varies from day to day, seasonally and by location as the atmospheric conditions that control it vary. Once the leaves of the crop cover the soil, ET will be close to ET_0 unless water is limiting, in which case growth and yield may be reduced. Water availability (precipitation, soil water storage and irrigation) determines how close ET will be to ET_0 and how close yield will be to the maximum possible yield.

Water availability

Maintaining an adequate supply of water to produce maximum grain yield requires matching the intermittent rainfall with the constant daily use of water by the crop (ET). The capacity of the soil to store water is the key to a successful match. The water for ET comes from the stored soil water, which is replenished by the proportion of rainfall that infiltrates into the soil. Successfully solving this intermittent supply–constant use dilemma is an important key to high grain yields; in fact, failure to solve it causes much of the year-to-year variation in rainfed grain yield. The successful solution and high grain yields are, more often than not, associated with a large storage capacity in the soil than with high rainfall.

Precipitation

The average annual precipitation is often used to characterize the water availability at a given location, but it is not a good indicator of the supply of water for crop growth, except at the extremes (obviously, the 254 mm (10 in) of annual rainfall in a desert is not enough to grow grain crops without irrigation). The distribution of precipitation throughout the year, especially that falling during the growing season, is a better indicator of the availability of water for crop growth. Distribution varies drastically from distinct wet and dry seasons found in tropical regions (e.g. the Brazilian Cerrado) or around the Mediterranean Sea or in California (Fig. 2.10) to locations with peak rainfall during the growing season, a nearly ideal distribution for grain crop production (Fort Dodge, Iowa; Fig. 2.10). Rainfall at many Midwestern and Southern US locations tends to decline during the growing season, reaching a minimum in the autumn (Lexington, Kentucky; Fig. 2.10).

Rainfall distribution is the key. Agriculture in California is completely dependent upon irrigation (summer rainfall is essentially zero; Fig. 2.10), allowing the producer complete control of water availability. The summer peak in Fort Dodge, Iowa, provides essentially the same growing season (June to September) rainfall as Florence, South Carolina (452 mm (17.8 in) and 493 mm

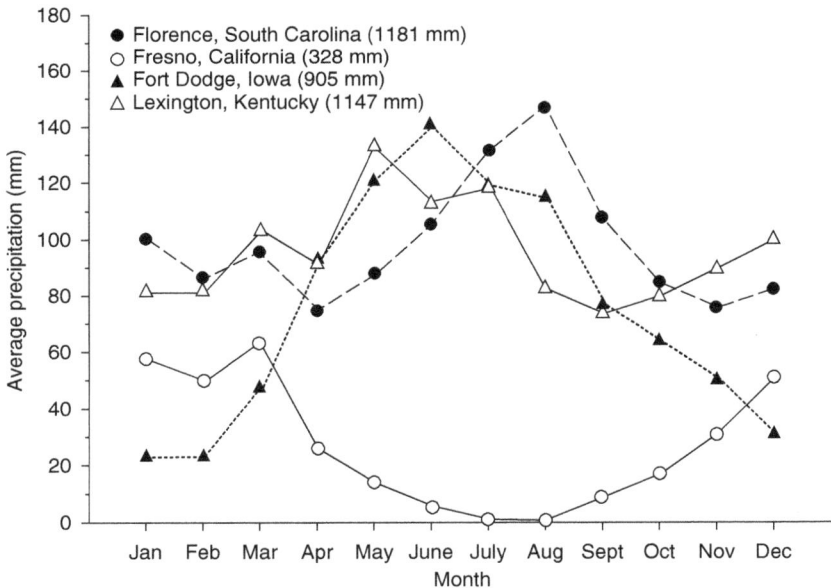

Fig. 2.10. Average monthly rainfall from four locations in the USA with large differences in distribution throughout the year. Yearly totals are given in the legend. Note: 25.4 mm = 1 in. (Data from the US Climate Data 2020, Version 30, National Weather Service.)

(19.4 in), respectively), but Florence has higher annual rainfall than Fort Dodge (1181 mm (46.5 in) versus 905 mm (35.5 in), respectively). When it rains during the year has a major impact on the type and productivity of agriculture in a region.

The growing-season precipitation, however, does not account for the distribution within the growing season, which is also important when matching supply with the demand for water. For example, 152 mm (6 in) of rain per month would be adequate for maximum yield in many climates if 38.1 mm (1.5 in) fell every week, but if 76.2 mm (3 in) fell on the first day and 76.2 mm (3 in) on the last day of the month, the crop could lose yield from drought stress. Two locations with the same monthly rainfall amounts could produce quite different yields depending upon the distribution within the growing season. The distribution within the growing season varies from year to year as a result of random variation in weather patterns that are not a function of climate. Consequently, the adequacy of rainfall in a cropping season at a specific location can only be assessed from data for that location in that season. This year-to-year variation in distribution encourages supplemental irrigation in climates with adequate average total growing-season precipitation to avoid stress during critical growth stages of the crop.

In summary, the ET_0, an estimate of the atmospheric demand for water, also plays an important role in determining the adequacy of the water supply as discussed previously. Cool, humid climates with lower ET_0 require less precipitation than climates with higher ET_0. The same amount and distribution of rainfall may produce maximum yield in a low ET_0 environment and only a modest yield in a high ET_0 environment. The ET_0 always sets the maximum water use and the supply, determined by rainfall (amount and distribution), the proportion of the rainfall that infiltrates into the soil and the size of the soil moisture storage container, must be adequate to meet this demand or yield will be reduced.

Infiltration

Rainfall reaching the soil surface has two options: it can run off, causing loss of water, soil and nutrients from erosion; or it can infiltrate into the soil where it may be available for crop growth, assuming it is not lost by deep drainage through the soil profile. Generally speaking, the infiltration rate will be highest on a coarse-textured soil (e.g. fine sandy loam), intermediate on a medium-textured soil (silt loam) and lowest on a fine-textured soil (silty clay loam). Soils on slopes have lower rates of infiltration than those that are flat. The infiltration rate is high when the soil surface is dry, but it declines rapidly after rainfall begins and the moisture content of the surface soil increases. These effects are modified somewhat by the condition of the soil surface.

Management practices that result in a rough soil surface or a surface covered with organic debris will slow the immediate runoff, giving water more time to infiltrate. The use of cover crops and/or no-tillage encourages infiltration and reduces runoff, thereby reducing erosion, a win–win situation (more water in the soil and less erosion). Producers cannot manage slope or soil texture, except by changing fields, but they can manage the condition of the soil surface to encourage infiltration. Managing surface conditions to enhance infiltration and reduce erosion will probably become more important in the future if, as predicted, the changing climate increases the frequency of high-intensity rainstorms.

Soil water storage

The amount of plant-available water held in the soil is primarily determined by soil texture and rooting depth. Soil water that is available for plant growth is the water held by the soil between the permanent wilting point (soil water level when the water is held so tightly by the soil particles that it is unavailable for plant growth) and field capacity (soil water level when deep drainage stops). The water held between field capacity and the permanent wilting point is the water that is available to the plant (plant-available water). Soil water levels at the permanent wilting point and field capacity are both affected by soil texture. A silt loam soil holds the most plant-available water per unit depth of soil (70.1 mmH_2O 30 cm^{-1} soil (2.76 in ft^{-1})) and coarse sands the least (12.1 mmH_2O 30 cm^{-1} soil (0.48 in ft^{-1})). Soil structure and perhaps the organic matter content also affect soil water-holding capacity, although the effect of organic matter levels is probably small. It is important to note that all of the plant-available water is not equally available for plant growth. Plants will experience water stress long before the soil water level is reduced to the permanent wilting point. Consequently, irrigation to avoid stress must start when some fraction of the plant-available water is depleted (often 60%). The determination of soil water levels at field capacity and the permanent wilting point is difficult and there is still considerable debate among soil physicists on how to define these critical levels of soil moisture (see the discussion in Scott, 2000, pp. 350–352). This debate does not negate the importance of the concept that there is a limit to the amount of plant-available water that can be held in a soil and this amount is a function of soil texture. Anyone farming a sandy soil is well aware of this limitation.

　　The depth of the root zone is often the primary factor determining the storage capacity of the soil, given that many agricultural soils are silt loams. Rooting depth, primarily a result of soil-forming processes, varies widely from the deep loess-derived soils in the Midwestern USA (rooting depth of 152 cm (5 ft)) to soils with compact layers (hardpans) that prevent root penetration and greatly restrict the rooting depth to as little as 38.1 cm (15 in). These restrictive layers can also be created by tillage operations in some soils. Deep tillage or subsoiling is sometimes used to destroy these hardpans, but it is rarely permanent.

　　These differences in rooting depth and soil water-holding capacity can have huge effects on yield. For example, the average soybean yield in Iowa from 2015 to 2019 was 3823 kg ha^{-1} (56.9 bu acre^{-1}) (NASS, 2020), which is nearly double that in South Carolina (2022 kg ha^{-1} (30.1 bu acre^{-1})). As noted previously, the average amounts of growing-season precipitation (June through September) in Iowa and South Carolina are similar, but the higher water-holding capacity of the deep loess soils in Iowa results in a closer match of supply and demand resulting in higher yield. The growing-season rainfall in South Carolina cannot overcome the low soil water-holding capacity (sandy coastal plain soils with a hardpan relatively near the surface) and yields are reduced. This comparison is probably influenced a little by higher ET$_0$ in South Carolina (higher temperatures) and potentially more disease and insect problems (also related to the higher temperatures), but the primary difference between the two states is the water-holding capacity of the soil. The water-holding capacity of the soil is a major determinant of productivity in rain-fed agriculture and is a character not readily increased by management.

In summary, water availability is a major determinant of crop yield. The variation of yield from year to year, within a field or region or between regions can often be related to the availability of water. Irrigation is one of the few management options a producer has to improve the match between the intermittent supply of water from rainfall and the constant demand determined by ET$_0$. The effective rainfall can be increased by maintaining organic debris on the surface via no-till, some form of conservation tillage or the use of cover crops to increase infiltration. Soil water-holding capacity is more or less fixed by soil characteristics, but it can be increased in soils with hardpans by mechanically breaking up the hardpan to increase rooting depth. It seems likely that managing the match between supply and demand will be more important in future climates as temperatures increase, rainfall patterns change and water availability for irrigation decreases.

Summary

We covered the fundamental processes of plant growth in this chapter, the processes that are ultimately responsible for the production of yield. Photosynthesis is the process that produces plant tissues using energy from the sun to incorporate CO_2 from the atmosphere into simple sugars. Respiration provides the energy to convert these simple sugars into all of the compounds making up the plant. Seeds, in particular, use sucrose and amino acids supplied by the mother plant to produce the oil, protein and complex carbohydrates that feed all of humankind. None of these processes can function without an adequate supply of water.

These are the basics, but to understand how plants produce yield we have to consider the functioning of these processes at the field or community level, because yield is produced by a community of plants, not by individual plants. The growth of crop communities is the focus of Chapter 3.

Growth of Crop Communities and the Production of Yield

3

No other occupation opens so wide a field for the profitable and agreeable combination of labor with cultivated thought as agriculture.

Abraham Lincoln (1809–1865), US President 1861–1865, established the USDA in 1862

Introduction

Chapter 2 focused on the fundamental plant processes that underlie the productivity of the crop community. The functioning of these processes determines yield; we cannot have high yields if environmental conditions (above- or below-ground), disease and insect pressures or management practices limit their operation. Understanding these fundamental processes is important, but it is not enough to provide a complete picture of the yield production process. A community of plants (a field of maize or wheat, for example) produces the economic yield of grain crops, so we must consider the growth of the community. Just as yield is measured on an area basis (kg ha^{-1} or bu acre^{-1}), we must consider photosynthesis and growth of the crop on an area basis. Many important processes that influence management decisions and yield operate at the community level; processes that regulate growth of isolated plants, on the other hand, often do not relate to growth of crop communities. Plant characteristics that influence yield of isolated plants may not be related to yield of plant communities.

Our goal in this chapter is to develop a general model of community growth and the production of yield that applies to all grain crops in spite of their many differences in morphology, growth patterns and reproductive characteristics. Such a model will provide a framework that will help us understand the yield production process, guide our thinking about the process and ultimately inform our management decisions leading to higher yields and efficient cropping systems.

© D.B. Egli 2021. *Applied Crop Physiology: Understanding the Fundamentals of Grain Crop Management* (D.B. Egli)
DOI: 10.1079/9781789245950.0003

The variation in reproductive morphology among grain crops complicates the development of a general model of community growth. Some species concentrate their seeds at a single location on the plant, while in other species the seeds are distributed over the entire plant. Soybean and other grain legumes (bean, pea, chickpea, broad bean, etc.) produce their seeds in pods that are borne at nodes on the main stem and branches. Consequently, the pods are distributed over the entire plant and each pod is relatively close to a leaf. The seeds of maize, on the other hand, are borne on a compact ear located near the middle of the main stem. Most modern maize hybrids produce only a single ear, but there are hybrids that produce more than one ear, especially when grown at low populations. Species that produce their seeds in a compact inflorescence at the top of the main stem or tillers include wheat, barley, rye (spike) sunflower (capitulum), rice and oat (branched panicle) and sorghum (compact panicle). Pods on groundnut (peanut) are produced in the soil at the end of the gynophore (peg). Canola or oilseed rape seeds are produced in pods (siliques) attached to the main stem. Interestingly, there is no clear evidence that reproductive morphology is, in any way, related to productivity or yield. We need a general growth staging system that can accommodate all of this diversity. Murata (1969) provided such a system.

Growth Staging Schemes

Fifty years ago, Murata (1969) (see also Egli, 2017, pp. 82–87) divided the yield production process into three stages:

I. Formation of organs for nutrient absorption and photosynthesis (vegetative growth).
II. Production of flower organs and the yield container (determination of seed number).
III. Production, accumulation and translocation of yield contents (seed filling).

Stage I represents vegetative growth, when the plant produces the leaves and roots that provide for and sustain community photosynthesis. Stage II represents flowering, pollination and the initial stages of seed growth and development. The number of seeds the crop will produce is determined during this stage. Stage III represents the seed-growth period, when the seeds accumulate the oil, protein and starch that give them their value. Yield is produced during this stage; Stages I and II are only preliminary events to the production of yield, the crop has not produced any yield at the end of Stage II. Stage III ends at physiological maturity, defined as the time when seeds reach their maximum dry weight (see Chapter 2, this volume, for a detailed discussion of physiological maturity).

Murata's three-stage system is very simple, but it is also very useful and helps us understand many facets of the yield production system. All grain crop species follow this pattern; first, the vegetative plant grows, and then it flowers and produces seeds, and finally the seeds grow to their maximum size and the plant matures. Nature, however, is never as simple as we would like it to be, so there are variations from this simple scheme. Stages I and II are not always entirely separate in time; for example, vegetative growth (Stage I) continues until the end of Stage II in soybean and probably other grain legumes, whereas Stages I and II are almost entirely separate in other species (e.g. maize). To complicate matters further, the beginning of Stage II is not clearly defined – does it begin when we see open flowers on the plant or when the flower primordia begin to develop before the flower opens? There is no clear answer to this question. The beginning of Stage II is often defined as when there are visual indications of flowering. Stress before flowering, however, can affect what happens during Stage II in some species, suggesting an earlier start to Stage II. These complications do not negate the value of this simple scheme and we will rely heavily on it to understand the yield production process and how to manage it to maximize yield and profits.

Important concepts embodied in Murata's scheme include:

1. The production of yield is a sequential process – first the crop produces the vegetative plant, then it flowers and sets seed (occurs simultaneously in some species), and finally the seeds are filled. Events occurring early in the sequence can affect growth later on, but, obviously, the reverse is not possible.
2. The three stages make it clear that the crop's growth activities vary throughout its life cycle. Consequently, the effect of an environmental event (occurrence of drought stress, for example) or a management practice on yield depends upon when it occurs or when it is applied. Timing is everything. A loss of leaves during Stage I may have no effect on yield, but the same loss during Stage II may cause significant reductions in the number of seeds and yield. Good managers are always aware of the growth stage of the crop.
3. Essentially no yield has been produced by the end of Stage II/the beginning of Stage III. Stages I and II are essential for the production of yield, but all of the yield is produced during Stage III. Most of the growth of the crop is devoted to preliminary activities and only a relatively small proportion is devoted to the actual production of yield. Seventy per cent of the total growth cycle of soybean (Maturity Groups I to IV) was devoted to Stages I and II and only 30% to Stage III (Egli and Bruening, 2000), making the length of Stage III a significant yield-limiting factor. Producing all of the yield in such a relatively short time puts a lot of stress on the productivity of the crop during Stage III. Devoting so much time to preliminary activities raises the question, is all that time needed or could it be reduced without affecting yield? We will discuss the importance of time in the production of yield in more detail later in this chapter.

More detailed growth staging systems are available for most grain crops (Table 3.1). These systems are crop specific and describe both vegetative and reproductive development, utilizing clearly defined, easily observable plant characteristics. Early descriptions of plant development often used subjective

Table 3.1. Sources of commonly used growth staging systems for grain crops.

Crop	Reference
Soybean	Fehr and Caviness (1977)
Maize	Hanway (1963); Abendroth *et al.* (2011)
Wheat and other cereals	Large (1954); Zadoks *et al.* (1974)
Sorghum	Vanderlip and Reeves (1972)
Groundnut (peanut)	Boote (1982)

descriptions that depended on the opinion of the individual making the rating. Modern systems avoid this problem by using clear definitions that are not subjective, such as measurements of size or counting leaves or nodes, etc. Several observers would arrive at the same growth stage for a plant when they are using a modern non-subjective system.

Growth staging systems facilitate communications within the grain crop community. The effect of the environment on crop yield or the timing of herbicide and pesticide applications is often growth stage specific, making clear, unambiguous communication between and among recommenders and producers extremely important. For example, the effect of a two-week period without rain could be catastrophic if it occurred during Stage II, but completely benign if it occurred during Stage I. Occasionally, researchers have changed definitions or added additional stages to established growth staging systems; this practice should not be undertaken lightly, because it will hamper communications (are you referring to the old definition or the new definition of that stage?). Most growth staging systems are based on easily observed plant characteristics; the relationships of growth stages to plant growth and the production of yield were developed later by crop physiologists and others studying these processes and their relationship to yield.

Growth staging systems usually define the growth stage of an individual plant. The growth stage of the community is the average of the individual plants staged (often ten plants); staging consecutive plants in a row usually provides a more representative sample than selecting 'normal' or random plants. Variation among plants or, more importantly, variation among locations in a field complicates making whole-field management decisions based on growth stages. Most management decisions do not require precision beyond the average for a field (e.g. a treatment that must be applied to soybean at growth stage R5 will probably not fail if the growth stage is R4.5 or 5.3). Sources for some of the commonly used growth staging systems are listed in Table 3.1. We will constantly refer to growth stages as we discuss crop management and yield production in the rest of this book.

Detailed growth staging systems are very crop specific, reflecting species differences in reproductive morphology and growth habit. However, Murata's three stages can be applied to all species, making it possible to develop a general representation or model of the yield production process that applies to all grain crop species. A unified model simplifies our discussion because we will not necessarily have to discuss each crop separately.

Growth of Crop Communities

Grain yield is produced by a crop community; consequently, the effects of management practices on productivity are often expressed at the community level. Our previous discussions of basic growth processes (Chapter 2, this volume) must be translated to the community before we can completely understand their role in the production of yield.

Murata's Stage I – vegetative growth

Growth of the vegetative plant during Murata's Stage I produces the roots and leaves – the photosynthetic factory – that will ultimately produce grain yield. Vegetative growth is the first part (if we ignore seed germination and seedling emergence) of the sequence that ends at physiological maturity. Vegetative growth is obviously an essential part to the yield production process, but, paradoxically, the rate of growth or the amount of vegetative growth is not always closely associated with yield. In 1971, Professor Bunting in England asked the question 'is your vegetative growth phase necessary?' and concluded 'yes, but not as necessary as one might think' (Bunting, 1971). As we shall see, this disconnect plays an important role in many management decisions. We will explore the logic behind Bunting's statement in this section of this chapter.

Figure 3.1 illustrates the accumulation of dry matter (growth) by a typical crop community. The cumulative dry weight of the community (leaves, stems and seeds) ($Mg\,ha^{-1}$ or $lb\,acre^{-1}$) increased steadily after seedling emergence until it reached a maximum at maturity. Day-to-day variation in environmental conditions (solar radiation, temperature, water availability) in the field could cause variation in photosynthesis and growth, resulting in a jagged curve instead of the smooth curve in Fig. 3.1b. Considering a smooth curve, however, makes it easier to understand the dynamics of community growth. The crop growth rate (CGR) is the rate of above-ground dry matter accumulation by the crop community (weight per unit area per unit time, usually expressed as $g\,m^{-2}\,day^{-1}$). The CGR increases slowly after the seedlings emerge from the soil, eventually reaching a constant rate before slowing down and stopping at maturity (Fig. 3.1b). Crop species with C_4 photosynthesis generally have higher CGRs than species with C_3 photosynthesis.

The growth depicted in Fig. 3.1 is a result of community photosynthesis, which is driven by the solar radiation absorbed by the leaves of the community. The absorbed solar radiation has two components: (i) the level incident on the crop is determined by location (latitude), elevation, time of the year and atmospheric conditions; and (ii) the proportion of the solar radiation intercepted by the leaves of the community, which is determined by the area of the leaves.

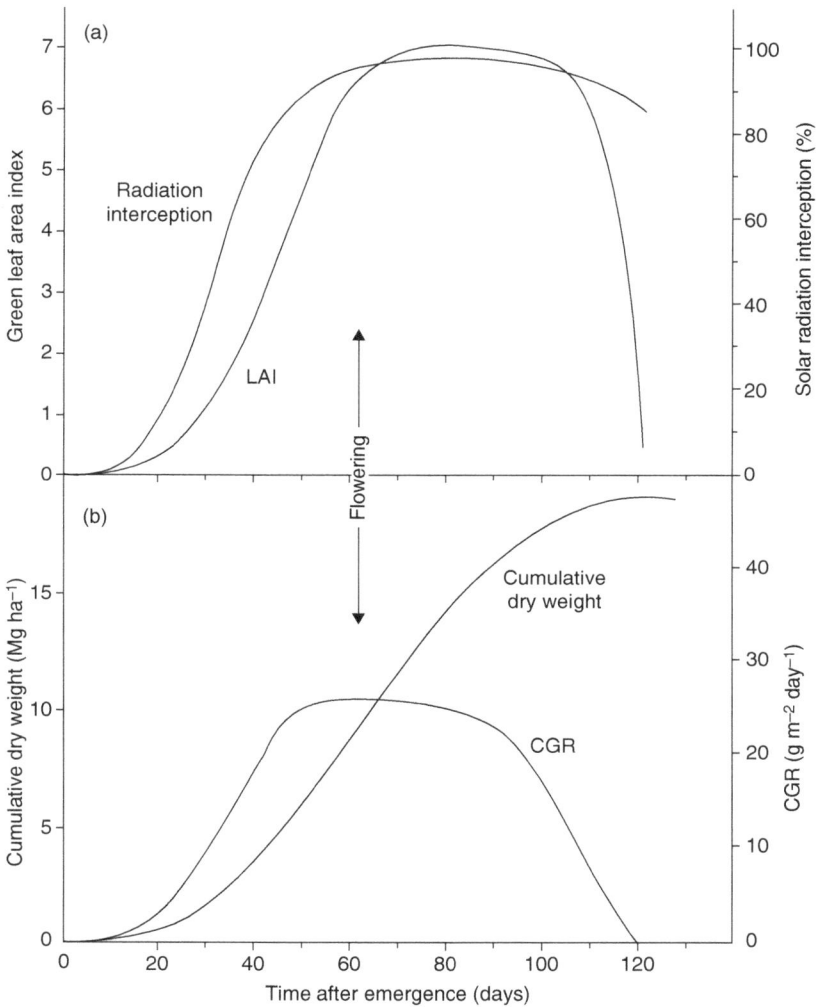

Fig. 3.1. Typical growth patterns of a grain crop in the field from emergence to maturity. Cumulative dry weight includes vegetative plant parts and seeds. CGR is the crop growth rate ($g\,m^{-2}\,day^{-1}$); LAI is the leaf area index. Cumulative dry weight is expressed in $Mg\,ha^{-1}$ (1 $Mg\,ha^{-1}$ = 1000 $kg\,ha^{-1}$ = 893 $lb\,acre^{-1}$). (Adapted from Gardner *et al.*, 1985. Used with permission from John Wiley and Sons.)

The area of the leaves produced by a crop community is usually described by the leaf area index (LAI), which is simply the total leaf area divided by the land area (Eqn 3.1):

Leaf area index (LAI) = leaf area/land area   (3.1)

LAI was first described by D.J. Watson, a famous crop physiologist in England in the middle of the last century (Watson, 1947). It expresses the idea that leaf area is only relevant when related to the land area occupied by the crop. For example, an LAI = 2.0 simply means that the crop community has 2.0 m² of leaves spread over 1.0 m² of land area, thus it provides an easy, dimensionless way to describe the leaf area of a crop. Leaf area is measured, traditionally, by harvesting all the leaves from a given area, measuring the area of one side of the leaf and dividing this total by the ground area of the sample. The orientation of the leaf (horizontal or vertical) has no effect on the LAI. Including only the green leaves (i.e. avoiding the senesced leaves) in the sample may provide a better estimate of the photosynthetic capacity of the community. The difference between total leaf area and green leaf area is probably small until leaves start senescing during seed filling (Murata's Stage III).

LAI is very small just after seedling emergence (Fig. 3.1a), but it increases rapidly to a maximum at the end of vegetative growth (Murata's Stage I) before declining as senesced leaves abscise. All grain crop species follow this general pattern, although the rate of increase and the maximum LAI will vary depending on species, variety, length of the growth cycle, management practices and environmental conditions. The magnitude of the decline in LAI as the crop matures (Fig. 3.1a) also varies among species, depending upon whether or not the senesced leaves fall off the plant (soybean is illustrated in Fig. 3.1) or if they remain on the plant (maize, wheat) and if the green leaf area or the total leaf area is measured.

The proportion of the incident solar radiation that is intercepted by the leaves (Eqn 3.2) determines the energy that is available for photosynthesis and transpiration:

$$SR_I = [(SR_A - SR_G)/SR_A] \times 100 \tag{3.2}$$

SR_I is the percentage of the solar radiation above the crop community (SR_A) that does not reach the soil surface (SR_G), i.e. it is intercepted by the leaves. The SR_I never reaches 100% because some of the solar radiation will always reach the soil surface (i.e. SR_G will never equal 0). Maximum interception is often taken as $SR_I = 95\%$. Equation 3.2 does not account for the incident solar radiation that is reflected by the leaves (i.e. SR_A should be reduced by the amount reflected). The reflection by most crop communities is roughly 10%, so not correcting for reflection does not create serious errors.

In the early stages of crop growth, the interception of solar radiation increases in step with the LAI until maximum interception occurs (~95% interception) (Fig. 3.1a), after which it stays constant until senescence and leaf abscission cause a decline. Intercepted solar radiation drives community photosynthesis, so maximum community photosynthesis and CGR occur only when the crop community has enough leaf area for maximum solar radiation interception as depicted in Fig. 3.2. The linear relationship between CGR and solar radiation interception (Fig. 3.2c) emphasizes the importance of interception in

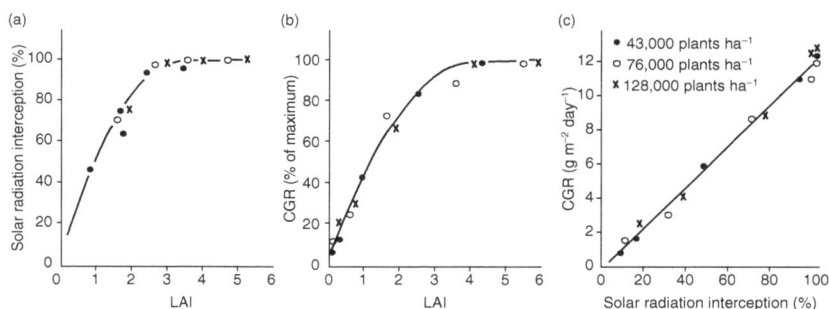

Fig. 3.2. Relationships between leaf area index (LAI), solar radiation interception and crop growth rate (CGR in g m^{-2} day^{-1}) in soybean. Plant populations of 43,000, 76,000 and 128,000 plants ha^{-1} (17,400, 30,769 and 51,822 plants acre^{-1}) were arranged in a hexagonal pattern. (From Gardner *et al.*, 1985, original research by Shibles and Weber, 1965. Used with permission from John Wiley and Sons.)

determining CGR. Solar radiation interception must reach maximum levels before or shortly after the beginning of Stage II or yield will be reduced.

The general shape of the LAI curve (Fig. 3.1a) is always the same, but the rate of increase in LAI after seedling emergence and the maximum LAI depend upon crop species, variety, management practices (planting date, row spacing and plant population) and environmental conditions. Leaf area will increase faster when crops are grown in narrow rows and at high populations and will usually reach the critical LAI (LAI producing 95% interception of solar radiation, LAI_{95}) sooner than crops grown in wide rows and low populations. In fact, the latter case could be so extreme that the crop would never reach the critical LAI by the end of vegetative growth.

Variety maturity influences the maximum LAI and when the crop reaches the critical LAI relative to the beginning of reproductive growth (Stage II). A longer growth cycle (later maturity) is usually associated with a longer vegetative growth period, resulting in a larger vegetative plant and a higher maximum LAI (Egli, 2011). The time between reaching the critical LAI and the beginning of reproductive growth increases as the length of the vegetative growth period increases (variety maturity is delayed). We will discuss the role of time and variety maturity in the determination of yield later in this chapter. Most modern grain cropping systems are managed to ensure that the critical LAI is reached well before flowering to avoid yield loss from incomplete solar radiation interception during reproductive growth.

The characteristics of the crop canopy (a species or variety characteristic), including the spatial arrangement and shape of the leaves, and leaf angle, also affect the relationship between LAI and solar radiation interception. Leaf angle, probably the most important characteristic, is a measure of the angle between the leaf blade and the horizontal. At the extremes, leaves are either horizontal (leaf angle = 0°) or vertical (leaf angle = 90°), but any angle in between is possible. Measuring the angle of all the leaves in a crop community is complicated, difficult and time-consuming for many crop species. We can, however, estimate the effect of leaf angle on the critical LAI using a theoretical equation based on the passage of solar radiation through the crop community (Table 3.2). Generally, the critical LAI for nearly vertical leaves (leaf angles approaching 90°) was higher than for horizontal leaves (leaf angle approaching 0°). Consequently, soybean with more horizontal leaves would have a lower critical LAI than maize with lots of vertical leaves. Exact critical LAI values would depend upon specific canopy characteristics of individual species and varieties.

At similar levels of solar radiation interception, vertical leaves may provide higher canopy photosynthesis and CGRs than horizontal leaves. Intercepted solar radiation is spread more evenly over the leaves in a community with vertical leaves (all of the leaves receive moderate levels of solar radiation) than it is in one with horizontal leaves (top leaves are exposed to high levels of solar radiation while leaves lower in the canopy are exposed to very low levels of solar radiation). The even distribution on the vertical leaves often results in higher canopy photosynthesis and higher CGRs.

The late Dr William G. Duncan, a remarkable Professor in the Agronomy Department at the University of Kentucky and a pioneer in the development of crop simulation models in the 1960s (Egli, 1991), used his original maize model to evaluate the potential value of vertical leaves. His simulations showed very clearly that vertical leaves increased canopy photosynthesis only when the LAI was greater than 3.0 (Fig. 3.3). The advantage over horizontal leaves increased as the LAI increased to 10.0. Increasing the leaf angle when the LAI was low, on the other hand, reduced community photosynthesis because solar radiation interception was reduced. The principle is clear: first, the community must maximize solar radiation interception and, when that is accomplished, intercepting it efficiently with vertical leaves will increase productivity. Dr Duncan's publication of this work (Duncan, 1971) represents one of the first publications in a refereed agronomic journal describing the use of a crop simulation model to evaluate a real-world crop management question. The model was useful in this case because it is very difficult to experimentally manipulate leaf angle in a field experiment that included the wide range in LAI

Table 3.2. Effect of leaf angle on the leaf area index (LAI) needed to intercept 95% of the incident solar radiation (the critical LAI, LAI_{95}).

Approximate leaf angle	Extinction coefficient	LAI_{95}[a]
Nearly horizontal	0.8	3.7
Intermediate	0.5	6.0
Nearly vertical	0.3	10.0

[a]Estimated from $I = I_0 e^{-kL}$, where I_0 is the solar radiation incident on the community, I is the solar radiation at the soil surface, k is the extinction coefficient and L is the leaf area index.

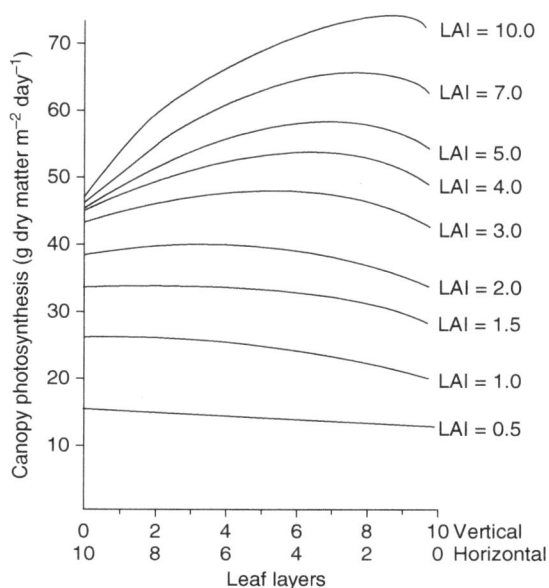

Fig. 3.3. The effect of leaf area index (LAI) and leaf angle on community photosynthesis of maize calculated with a crop simulation model. The community was divided into ten layers of leaves and all of the leaves in a layer were either vertical or horizontal. Vertical leaf layers were on top of horizontal leaf layers. (From Duncan, 1971. Used with permission from John Wiley and Sons.)

needed to evaluate the interaction of leaf angle and LAI. Once the principle was established, subsequent experimental work verified the predictions of the model (Winter and Ohlrogge, 1973) and led to the development of maize hybrids with more vertical leaves that contributed to the historical increase in maize yield (Duvick, 2005). Crop simulation models are widely used today to investigate the physiological basis of crop yield, to analyse long-term effects of cropping systems and the potential impact of climate change.

ET by a crop community generally follows the LAI curve in Fig. 3.1a. The soil surface is usually dry, so soil evaporation is low (as discussed in Chapter 2, this volume), limiting ET when the LAI does not provide complete ground cover. If the soil surface is wet (just after a rain, for example), soil evaporation will be high and ET will be high until the soil surface dries. When the leaves completely cover the soil surface (LAI ≥ LAI_{95}) most of the ET comes from transpiration which has access to water below the soil surface and ET reaches a maximum for that particular environment (assuming water is not limiting).

The relationships of LAI, solar radiation interception and CGR summarized in Fig. 3.2 have a number of important implications for crop growth and the production of yield; implications that help us manage our crops for maximum yield.

1. The maximum CGR occurs when solar radiation interception reaches a maximum (Fig. 3.2c). Maximum CGR will produce maximum yield, so management systems must be designed to produce complete solar radiation interception by the beginning of reproductive growth – anything less will reduce yield. Solar radiation that is not intercepted is wasted; actually, it is worse than wasted because it will help weeds grow.

2. Increasing LAI above the level providing maximum solar radiation interception does not increase CGR (Fig. 3.2b). If CGR is not increased, there is no reason to expect yield to increase. Consequently, varieties that produce large vegetative weights and high LAIs (i.e. late-maturing varieties) do not necessarily produce higher yield. There are, however, some advantages to having an LAI above the critical level. First, defoliation by insect feeding will not affect yield until the LAI is reduced below the critical level. So the 'excess' LAI provides some protection from defoliating insects. In fact, the potential effects of insect feeding on yield should be judged by the effect it has on solar radiation interception, not the reduction in LAI. Excess LAI also provides protection against transient environmental stress that reduces vegetative growth and LAI. The stress may not affect yield (assuming the stress is relieved during reproductive growth) if the LAI after the stress is still above the critical level. Yield will be reduced, however, if the LAI is below the critical level during reproductive growth. Arranging a crop production system to just produce the critical LAI by the beginning of reproductive growth (e.g. by reducing the plant population) will produce maximum yield if there is no stress; excess LAI provides some protection against loss of LAI from environmental stress during vegetative growth. There is little evidence for an optimum LAI (CGR decreases as LAI increases above the level producing the maximum CGR) in grain crops. Managing crops to produce an optimum LAI would be difficult; much harder than just making sure the LAI is at or above the level producing 95% solar radiation interception (the critical level).

3. Narrow row spacings, high plant populations and late-maturing varieties (long vegetative growth period) can be used in many crops to produce the critical LAI early in vegetative growth, well before the beginning of reproductive growth. Does this early occurrence have any effect on crop productivity? Considering only CGR, the answer is no. Maximizing CGR during reproductive growth is all that is needed to maximize yield. It is true that the crop will intercept more total solar radiation over its life cycle when the critical LAI occurs early in vegetative growth. Despite claims to the contrary, radiation intercepted before the beginning of reproductive growth seldom contributes directly to yield. The question, however, is more complicated than just the interception–CGR relationship. Reaching the critical LAI early in vegetative growth, well before flowering, will increase water use during vegetative growth because ET by the crop community reaches a maximum when LAI reaches the critical LAI. If water becomes limiting later during reproductive growth, the excess early use and consequent depletion of the soil water supply could result in the earlier initiation of water stress during reproductive growth. If water is not limiting, the early increase in LAI will be, from a yield viewpoint, neutral. On the other hand, management practices (e.g. wide row spacings) that delay the occurrence of the critical LAI will reduce water use during vegetative growth, 'saving' it for the reproductive growth phase. This tactic is useful in climates where grain crops normally experience water stress during seed filling; the water 'saved' will delay the occurrence of stress and increase yield. This approach does not lead to maximum yields; it is just making the best of a bad situation in water-limited environments.

Early development of the critical LAI will inhibit weed growth, surely a potential advantage in many cropping systems. Solar radiation intercepted by the crop is not available for weed growth, so early closure may increase yield (or at least reduce herbicide costs). The advantages from a weed control standpoint will probably outweigh the potential negative effects from a water use perspective in humid environments.

In summary, the effect of LAI on solar radiation interception plays an important role in managing grain crops for maximum yield. CGR is directly related to the level of intercepted solar radiation, so ensuring maximum interception by the beginning of reproductive growth is necessary for high yield. It does not, however, guarantee high yield; other aspects of the environment also influence CGR and yield. Not reaching the critical LAI by the beginning of reproductive growth, however, will mean that yields are reduced.

Reproductive growth

Yield components

Relating yield to specific plant growth processes is often difficult because plants do not really 'produce' yield. Yield is a construct developed by humans to measure plant productivity. Plants produce flowers, flowers pollinate, seeds are formed and then they grow until they reach maturity. At that time, humans measure the collective weight of the seeds per unit area and call it yield. To relate yield to the plant processes that produce it, we have to consider flowers, seeds and their growth. In other words, we have to break yield down into its components – we must think about yield components, not yield per se. Engledow and Wadham (1923) may have been the first to use the yield components approach in their analyses of the response of wheat to changes in plant population in 1923 (Evans, 1993, p. 260).

Evaluating yield components instead of yield often helps us understand the effect of the environment or management practices on yield. Yield is the final product of many environmentally sensitive morphological and physiological processes integrated over the 100 to 120 days or so from planting to maturity. Integration of these processes over time creates many opportunities for the environment to affect growth and ultimately yield. The complexity of the growth stage by environment interaction is not obvious when evaluating yield, but some of it can be unravelled by dividing yield into its components.

The simplest expression of yield components divides yield into seed number (seeds area^{-1}) and seed size (weight seed^{-1}) (Eqn 3.3):

Yield (weight area^{-1}) = seed number (seeds area^{-1}) × seed

size (weight seed^{-1})   (3.3)

Much more complicated expressions can be written (e.g. Eqn 3.4 for soybean)

Yield (weight area^{-1}) = (plants area^{-1}) (pods plant^{-1})

(seeds pod^{-1}) (weight seed^{-1})   (3.4)

but creating more components does not necessarily help us understand the yield production process; in fact, the extra complexity may create more confusion than clarity. Equation 3.4 is an accurate representation of soybean yield, but it contains plant population (plants area^{-1}) and a term that is affected by plant population (pods plant^{-1}). Increasing soybean population over a substantial range (see Chapter 4, this volume) is usually associated with a decrease in pods per plant with no change in yield. The variation in pods per plant is not necessarily associated with yield, so Eqn 3.4 tends to confuse rather than clarify the yield production process. Equation 3.4 is also species specific, reflecting the reproductive morphology of individual species that was discussed at the beginning of this chapter. Consequently, each crop will have its own detailed equation, making it difficult to compare crop species. Equation 3.3 applies to all grain crop species, making it possible to develop a unified description of the yield production process that applies to all species. Collecting data representing the four terms in Eqn 3.4 or for detailed equations for other crop species requires much more effort than the two terms in Eqn 3.3. Evaluation of the two terms in Eqn 3.3 requires only measuring seed size (weight per seed) and then dividing yield by seed size to get seed number.

The most important attributes of Eqn 3.3 are: (i) it describes yield for all grain crop species; and (ii) it relates directly to Murata's stages of development – seed number is determined during Stage II and seed size is determined during Stage III. Combining Murata's stages of yield production and the simple yield component equation gives us a simple, but powerful, model of the yield production process in all grain crop species, regardless of their reproductive characteristics.

Valid estimates of yield components in the field require samples that accurately reflect the yield characteristics of the crop community. Problems can arise when individual plants are used to estimate the components. A common practice is to select 'typical' or 'random' plants from the community to measure components, but yield calculated from these estimates is often substantially higher than yield measured by harvesting all the plants in a specific area. The 'typical' or 'random' plants were not representative of the plant community because only large plants were included in the sample, small plants were ignored. Yield components from a non-representative sample cannot provide any useful information about the yield production process. A better approach is to sample all the plants (large and small) in a specific area or a given length of row. Estimating seed size from the seed sample used to determine yield and calculating seed number (yield/seed size) provides a representative estimate of the components in Eqn 3.3.

Yield components are most useful to understand the effect of the environment or management practices on yield. Yield components help answer questions such as 'why did management practice X increase yield?' or 'why was the yield in environment I less than in environment II?'. The yield component that is responsible for variation in yield identifies the growth stage that was important in determining yield. If a yield-changing management practice increased seed number, it must have affected crop growth during Stage II. Conversely, a change in seed size signifies that Stage III was affected. Evaluating yield components provides information about the production of yield; information that helps us understand why and how a management practice or the environment affected yield.

Yield components that are under genetic control are not as useful, because they may not be related to yield. The yield components of the two soybean varieties in Table 3.3 were not related to yield. These varieties with significant differences in seed number and seed size produced the same yield. Seed size in these two varieties was under genetic control, so the small-seeded variety produced more seeds to compensate for the small seeds. This is an example of the dreaded 'yield component compensation' – change one component and another will change in the opposite direction and yield remains constant – that caused plant breeders to abandon the use of yield components as selection criteria. One important lesson from this example is that large-seeded varieties do not necessarily produce higher yield. The compensation of size and number is more dramatic when comparing species (Table 3.4). Rice and maize both produce high yields, but the small-seeded rice produces roughly ten times as many seeds as the large-seeded maize. Soybean produced slightly higher yield than wheat with nearly 80% fewer seeds. Genetic differences in seed size provide no

Table 3.3. Yield and yield components (seed number and seed size) of two soybean varieties with genetic differences in seed size grown in the same environment. (Adapted from Egli, 1993a.)

Variety	Yield[a] $(g\,m^{-2})$	Seed number[b] $(seeds\,m^{-2})$	Seed size[c] $(mg\,seed^{-1})$
'Harper'	337	1668	200
'Essex'	330	2156	152
	(NS)	*	*

*Significant at $\alpha = 0.05$; NS = not significant.
[a] $g\,m^{-2} \times 0.1488 = bu\,acre^{-1}$.
[b] $Seeds\,m^{-2} \times 4048.6 = seeds\,acre^{-1}$.
[c] $Seeds\,lb^{-1} = (453.6)/[(mg\,seed^{-1})(0.001)]$.

Table 3.4. Species differences in seed size, yield and seed number. (From Egli, 2017, p. 90.)

Species	Approximate seed size[a] (mg seed^{-1})	Average yield[b] (g m^{-2})	Seed number[c] (seeds m^{-2})
Rice	28	849	30,321
Wheat[d]	41	296	7,220
Sorghum	28	424	15,143
Maize	302	1,039	3,340
Soybean	202	312	1,544
Bean[e]	345	201	583

[a]From Table 2.2.
[b]Average US yield for 2013, 2014 and 2015. Data from National Agricultural Statistics Service (NASS, 2020).
[c]Seed number = yield/seed size. See footnotes to Table 3.3 for conversions to bu acre^{-1}, seeds acre^{-1} and seeds lb^{-1}.
[d]Winter wheat.
[e]*Phaseolus vulgarius* L.

information about the yield potential of a variety or species. We will discuss this relationship in more detail later in this chapter.

Murata's Stage II – seeds per unit area (sink size)

We can see from the yield component equation (Eqn 3.3) that high yield can be associated with many seeds, large seeds (non-genetic variation in size) or any combination thereof. Soybean yield, however, was closely associated with variation in the number of seeds per unit area (Fig. 3.4). Seed size (weight per seed) varied, but it was not associated with yield. Wheat (Fig. 3.5) and maize (Egli, 2017, p. 90) show the same relationship between seed number and yield as soybean, as would all other grain crops. High yield in these crops was a result of large numbers of seeds per unit area; seed size did not contribute to the environmentally induced variation in yield.

Seed number in these examples (Figs 3.4 and 3.5) was responding to variation in the productivity of the environment. Genetic differences in crop productivity would also be associated with seed number. Differences in yield associated with seed-fill duration, however, would not be associated with seed number.

What mechanism or plant process gives seed number its pre-eminent position? It is very simple: seed number is determined first. Murata's Stage II (formation of flower organs and the yield container) (Murata, 1969) occurs before Stage III (seed filling). Stage II represents the first opportunity the crop has to adjust its reproductive output to its productivity level; this adjustment produces the close association between productivity (determined by environmental conditions, species, variety and management practices) and seed number. Being first makes seed number much more important than seed size as a yield determinant.

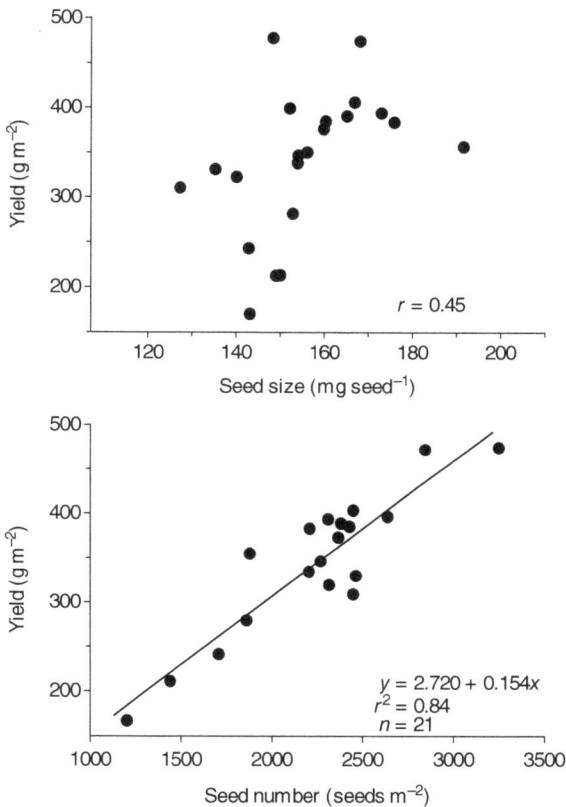

Fig. 3.4. The relationship between soybean yield and the yield components seed size (mg seed[-1]) and seed number (seeds m[-2]). The variety 'Iroquois' (Maturity Group III) was grown at 21 locations in 1996. (Unpublished data from the Uniform Soybean Test – Northern Region. From Egli, 2017, p. 88.)

It is not easy to relate the definition of Murata's Stage II to growth stages of individual crops as discussed previously. We can simplify this problem by thinking of Stage II as the 'critical period' for seed number determination, which is defined as the period when seed number is sensitive to environmental conditions. Stress that reduces photosynthesis during the critical period will reduce seed number; stress occurring before or after the critical period will have minimal effects on seed number.

The critical period for soybean is from growth stage R1 (initial bloom) to between growth stages R5 and R6. The critical period for maize is often taken as the period from 10 to 15 days before silking to 20 days after silking. Seed number in wheat seems to be sensitive to the environment from 20 days

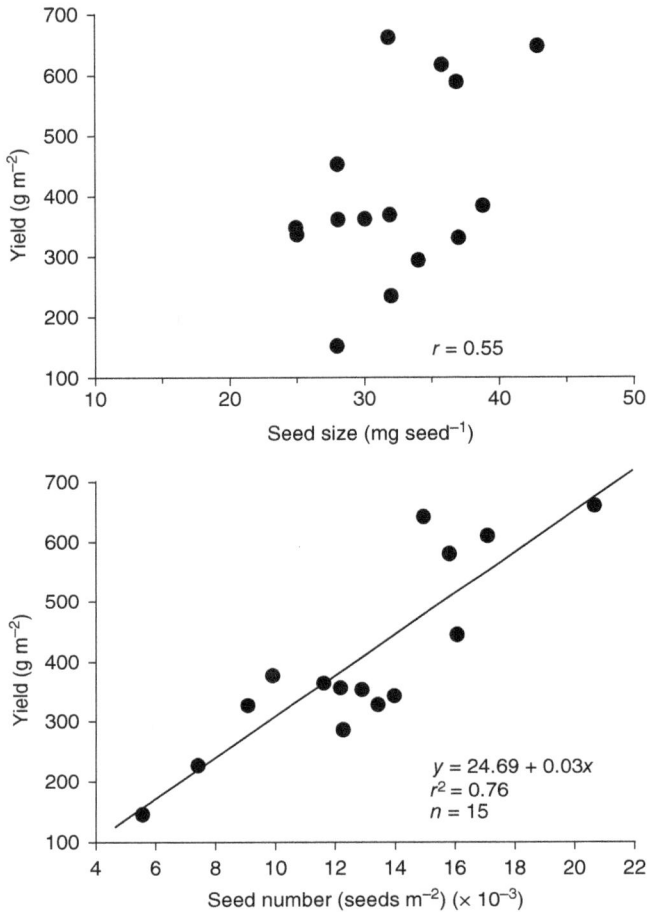

Fig. 3.5. The relationship between wheat yield and the yield components seed size (mg seed⁻¹) and seed number (seeds m⁻²). The variety 'Cardinal' was grown for 15 location-years. (Unpublished data from the 1990/91 Uniform Eastern Soft Red Winter Wheat Nursery Program. From Egli, 2017, p. 89.)

before to 10 days after anthesis (Egli, 2017, pp. 84–86). The critical period has been defined for many grain crop species, but it must be noted that the critical period cannot be defined precisely (i.e. seed number will be affected by the environment today but not tomorrow), rather it is a general guide to when seed number is sensitive to environmental conditions. Environmental conditions before the critical period can indirectly influence seed number; for example, if stress during vegetative growth, before the critical period, reduces the LAI, solar radiation interception and community photosynthesis during the critical period, seed number will be reduced. The extent that environmental conditions during the pre-critical period influences the development of reproductive structures and ultimately seed number is not clear.

High yields require large numbers of seeds (Figs 3.4 and 3.5), making the critical period an extremely important growth stage for the determination of yield. The crop growth rate (CGR) is a measure of the productivity of the crop, so it is not surprising that there is a close association between seed number and CGR during the critical period (Fig. 3.6). Reducing CGR in Fig. 3.6 by placing varying levels of shade cloth over the crop community reduced seed number of both soybean varieties. Seed number was related to the rate of photosynthesis and the growth rate of the community (i.e. CGR) across 3 years. Research has documented similar relationships for other grain crops.

Relating seed number to CGR and the availability of assimilate from photosynthesis establishes a powerful link between the productivity of the crop

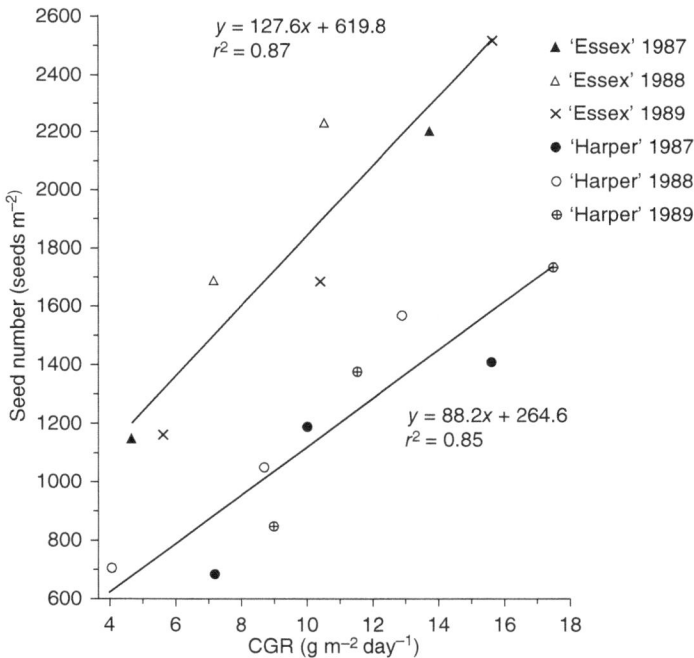

Fig. 3.6. The relationship between seed number (seeds m^{-2}) and crop growth rate (CGR) for two soybean varieties that differ in seed size and seed growth rate (SGR). 'Essex' produces small seeds (averages 147 mg seed^{-1} or 3088 seeds lb^{-1}) with a mean individual SGR of 4.5 mg seed^{-1} day^{-1}. 'Harper' produces larger seeds (averages 193 mg seed^{-1} or 2352 seeds lb^{-1}) with a higher individual SGR (6.3 mg seed^{-1} day^{-1}). (Adapted from Egli and Yu, 1991. Used with permission from John Wiley and Sons.)

and its reproductive potential (i.e. seed number). Such a relationship makes it possible for the crop to maximize its reproductive output for any level of productivity. In evolutionary terms, relating seed number to CGR maximizes reproductive fitness and the chances of survival of viable offspring; from the viewpoint of a modern grain producer, this relationship maximizes yield potential in any environment.

The adjustment of seed number to CGR during Stage II makes this stage a critical period in the crop's life cycle. Any stress that reduces CGR during this period, whether it is nutrient stress, a lack of water, reduced solar radiation (Fig. 3.6), insect or disease pressure, or failure of the community to maximize solar radiation interception, will reduce CGR and seed number. A transient stress (a dry period, for example) during Stage II reduces seed number regardless of the productivity of the crop before or after Stage II. The crop, depending on the species, may be able to recover some of the 'lost yield' by increasing seed size if environmental conditions improve after Stage II. We will discuss this aspect of the yield production process later in this chapter when we cover seed filling (Stage III).

How does the plant adjust the number of seeds it produces to bring it in balance with the productivity of the environment? The mechanisms are very species specific, determined by the reproductive morphology of the crop. The adjustment usually occurs in two phases: (i) the crop produces a large number of potential seeds (flowers); and (ii) this potential seed number is adjusted downward to match the productivity of the crop. It seems that the crop starts out wildly optimistic about its potential reproductive output, but then, facing reality, the potential is adjusted downward to match the productivity level.

Increases in potential seed number per plant occur when plants produce seed-bearing tillers (wheat, barley, rice and sorghum), branches (soybean and other grain legumes, canola (oilseed rape)) or increase the number of flower-producing nodes on the main stem (soybean and other grain legumes). Soybean (and presumably other grain legumes) can also increase the number of flowers per node. These characteristics make a plant 'plastic', i.e. it has the ability to increase seeds per plant in response to improved environmental conditions. Plastic species can easily capture an increase in the productivity of the environment by increasing seed number. Species that lack plasticity cannot easily increase seeds per plant. Plant breeders reduced the plasticity of maize to the point that an individual plant has a very limited capacity to respond to an improved environment. For example, modern maize hybrids rarely produce ear-bearing tillers and many produce only a single ear. Multiple-ear types exist but they are usually not favoured by commercial breeders or producers. Variation in the number of florets per ear would add some plasticity to the maize plant, but there is not much evidence supporting the existence of this 'flex' characteristic. Sunflower also has very little plasticity.

Plasticity plays an important role in the response of a crop to changes in plant density. Plastic species usually produce the same yield over a substantial range in population: as population decreases, reproductive output per plant increases, maintaining a constant yield. Non-plastic species are much more sensitive to changes in population because the capacity to increase reproductive output per plant as population decreases is limited. We will discuss population responses in detail in Chapter 4 (this volume).

The second phase of the adjustment process occurs when potential seed number is reduced until it matches the capacity of the plant to fill the seeds. Reductions occur when tillers fail to produce reproductive structures or flowers, or developing pods or seeds abort. This downward adjustment almost always occurs, even in highly productive environments. For example, abortion of flowers and small pods in soybean never approaches zero; in fact, 50% abortion was measured in high-yield environments. The number of competent florets in wheat was 60 to 70% less than the number of flower primordia (Egli, 2017, pp. 92–96).

Developing a larger sink (potential seed number) than needed and adjusting it downward helps ensure that yield will not be limited by the potential number of seeds; instead, it is limited by photosynthesis of the crop community during the critical period. Non-plastic species, such as maize, represent an exception to this rule; seed number can be limited by the size of the ear. If the ear doesn't produce enough flowers (potential seeds), seed number will limit the conversion of the available photosynthate into yield. The solution to this problem in maize is to increase the plant population, thereby increasing the number of ears and potential seeds per unit area.

Reproductive failure, whether it is flower or pod abortion in soybean, failure of maize to fill the tip kernels on the ear, death of developing tillers in wheat and other cereals, or any other form of reproductive abortion, is often considered a negative factor leading to 'lost' yield. The 'lost' yield syndrome is particularly common among producers of soybean (with its high level of flower and small pod abortion) and maize (with its highly visual failure of ears to fill to the tip) and it is expressed as 'if those flowers did not abort, yield would be increased'. This approach is completely wrong; abortion occurred to match seed number to the assimilate supply from photosynthesis and it does not represent, by itself, lost yield. There was no effect on yield when researchers used growth regulators to artificially increase seed number by decreasing abortion without an increase in assimilate supply. A decrease in seed size compensated for the increase in seed number with no change in yield.

Reproductive failure is a normal part of the adjustment of seed number to the assimilate supply. Excessive abortion is an indication that the assimilate supply has been reduced and that reduction should be the point of concern, not the abortion process. In fact, one can argue that abortion is good because it is an indication that yield was not limited by the number of flowers or seeds (i.e. a sink limitation); or, in other words, there were more than enough flowers and seeds to utilize all of the available assimilate. A maize ear that is filled to the very tip raises the question 'would there be more seeds and higher yield if the plant had a bigger ear or if there were more ears per unit area (the population was higher)?', i.e. was the crop sink limited? Aborted kernels at the tip of the ear indicate that there was no sink limitation and yield utilized all of the available photosynthate. The aborted kernels could also indicate a stress-induced reduction in the assimilate supply. Ideally, the population should be high enough so that all of the flowers are not needed to produce maximum yield.

Each crop has several mechanisms by which it can increase or decrease the number of seeds it produces. These mechanisms balance the number of seeds that develop with the ability of the crop to fill the seeds, thereby maximizing seed number for any level of productivity. Producing too many seeds results in smaller than normal seeds, while producing too few seeds may result in a sink limitation and reduced yield. Grain crops have evolved to reach the correct balance, but there are two important exceptions to this rule. First, seed number is in balance with the CGR during the critical period (Stage II), but after the critical period environmental conditions and the CGR can change during seed filling (e.g. rain occurs during seed filling, relieving a drought that occurred during Stage II) creating an unbalanced situation (seed number was set too low). The crop is no better at forecasting future weather conditions than humans. We will discuss this interaction at length when we discuss Stage III (seed filling).

Second, certain environmental stresses may affect seed number independently of CGR and crop productivity, thereby preventing the establishment of the correct ratio and the production of normal-sized seeds. Excessively high temperatures during the critical period (Murata's Stage II) may disrupt pollination or fertilization in maize, soybean, rice and cowpea, as can B deficiency in wheat or low temperatures in rice (Egli, 2017, p. 96). N deficiency or moisture stress can cause pollen shed in maize to occur before the silks appear, thereby preventing pollination (Egli, 2017, p. 96). These mechanisms reduce seed number and disrupt the relationship between seed number and the assimilate supply. Fortunately, these disruptions are usually not common or widespread, but they can cause catastrophic reductions in seed number and yield when they occur.

The length of the critical period (Stage II) varies among crop species and among varieties within a species. Stage II in soybean, for example, extends from growth stage R1 until somewhere between growth stages R5 and R6, which can occur 40 or more days later, while maize has a much shorter critical period (~28 days), as do other cereals. Variation among varieties is often related to maturity of the variety; as maturity is delayed, the length of the total growth cycle and Stage II increases. Stage II in soybean increased by 12 days (46%) from Maturity Group I to IV while the total growth cycle increased by 29 days (30%) (Egli and Bruening, 2000). Delayed planting may decrease the length of the critical period.

Some scientists argue that a longer Stage II provides more days for the crop to accumulate assimilate from photosynthesis, which will translate into more seeds. Other scientists argue that relating seed number to the length of Stage II uncouples seed number from the productivity of the crop, expressed through the CGR, and disturbs the balance between seed number and the capacity of the crop to fill the seeds. Relating seed number to duration associates it with an accumulated total over a period of time, not a daily supply or rate as embodied in the CGR. Seed number based on duration could be set well above the number that could be supported by the CGR during seed filling, resulting in smaller seeds and possible reductions in market quality. Some of our most productive species have relatively short critical periods (e.g. wheat), suggesting that long critical periods are not necessary for large numbers of seeds and high yield. From a theoretical perspective, it seems unlikely, in my opinion, that the duration of Stage II is an important determinant of seed number in grain crops.

Crops with a shorter Stage II may be less tolerant of short periods of stress (1 or 2 weeks of dry weather, for example) during this stage, suffering larger reductions in seed number than crops with a longer Stage II. A longer Stage II may allow the crop to recover from the stress before the end of the stage, thereby minimizing reductions in seed number. Experiments with soybean, however, did not support this thesis. Reductions in photosynthesis for just the first 14 days of Stage II reduced seed number; the crop could not recover during the remaining 29 days of Stage II when photosynthesis was restored (Egli, 2010), so the long Stage II was of little value. The longer duration of Stage II showed some value when soybean yield (and presumably seed number) was less variable across years than maize yield in seven of 13 comparisons of long-term (8- to 20-year) crop rotation studies (Egli, 2017, p. 127).

Overall, the data supporting the value of a long Stage II are not very strong. Neither theoretical considerations nor species comparisons provide much support. Relating seed number to the CGR during Stage II, without considering its duration or the remobilization of storage reserves, seems to provide the best explanation of how grain crops determine the number of seeds to produce.

We discussed the relationship between crop productivity and seed number at some length, but we have not discussed the relationship of seed number to genetic differences in seed size (weight per seed). As seed size decreases, seed number increases, but in this case, yield is not associated with seed number. This relationship is obvious in any comparison of crop species. Rice has small seeds and a somewhat lower average yield than maize (Table 3.4) but produces roughly ten times as many seeds. Wheat (small seeds) and soybean (large seeds) produced comparable yields, but wheat produced five times as many seeds. The soybean variety 'Essex' (small seeds) (Fig. 3.6) produced more seeds at a common CGR than 'Harper' (large seeds), but the two varieties produced the same average yield (Table 3.3) because seed number compensated for seed size. We can understand this difference by considering the mechanisms responsible for variation in seed number.

The two varieties in Fig. 3.6 exhibit genetic variation in seed size and, more specifically, genetic variation in SGR (see Chapter 2, this volume), which is controlled by the seed. Small seeds with low SGRs require less assimilate to support the growth of a single seed than large seeds with high SGRs. Consequently, a given supply of assimilate from photosynthesis can support more slow-growing seeds (i.e. 'Essex', Fig. 3.6) than fast-growing seeds (i.e. 'Harper', Fig. 3.6), so

'Essex' produced more seeds than 'Harper' at the same CGR. This relationship partially explains why rice (small seed, low SGR) produced ten times as many seeds as the large-seeded (high SGR) maize (Table 3.4). Differences in assimilate supply in this example could also influence seed number.

The variation in seed number related to genetic variation in seed size and seed growth rate is not related to yield. Seed number determined by genetic differences in seed growth rate has nothing to do with productivity of the crop – the available assimilate, determined by the productivity level, is simply packaged differently, i.e. in more small or fewer large seeds. The seed number–yield relationship occurs when seed number is related to the productivity of the crop and the availability of assimilate (CGR); this relationship is completely uncoupled by genetic variation in SGR, so the relationship with yield is lost. Genetic variation in seed size is usually closely associated with SGR, so genetic variation in size is not related to yield.

There is one exception to this rule. If genetic differences in seed size are determined by seed-fill duration, not SGR, large seeds will produce higher yield. Species with seeds that are large because they grow for a longer time will, with all other factors equal, yield more than species with small seeds associated with short seed-filling periods. Varieties with the large seed–long seed-filling period combination exist, but they are extremely rare, so assuming that genetically large seeds have high SGRs and will not produce higher yields is the safe bet.

When soybean breeders first started developing varieties with improved yield potential in the early years of the last century, it seemed obvious to some that selecting for large seeds would lead to higher yield. They were successful – seed size increased, but, unfortunately, yield did not change. The increase in size was compensated for by a decrease in seed number and yield stayed the same, providing a classic example of 'yield component compensation' – a change in one yield component is compensated for by an opposite change in another yield component to maintain a constant yield. Yield component compensation makes it impossible to use seed size as an indicator of a high-yielding variety or species. Genetic differences in seed size are not usually related to yield.

In summary, all grain crop species have multiple mechanisms to adjust the number of seeds that they produce to the productivity of the environment. These adjustments occur in the early stages of reproductive growth (Murata's Stage II), a critical period in the yield production process. Evolution favoured these mechanisms because the adjustment maximized the number of seeds in any environment, which maximized reproductive fitness and survival of the next generation. By favouring these mechanisms, evolution also produced a plant well suited to respond to the productivity of any environment and maximize yield. Selection for yield by humans minimized these adjustment mechanisms in some species (principally maize and sunflower), requiring management changes by producers (usually increasing plant population) to avoid sink limitations (not enough flowers and seeds) and reduced yield. Stress-induced reductions of seed number will usually result in lower yields, given the limited ability of most crop species to compensate by increasing seed size. Consequently, there is no doubt that Stage II is the critical stage in the production of yield.

Murata's Stage III – seed filling and seed size

Yield is essentially zero at the beginning of Stage III (the total weight of the tiny undeveloped seeds is negligible relative to their weight at maturity), but all of the preliminary events are finished. The vegetative plant has stopped growing; the production of leaves that will produce the photosynthate to power yield production and the roots that will support it are finished. The number of seeds is fixed, and the tiny immature seeds are ready to start growing. Stage III represents the main event – when the production, accumulation and translocation of yield contents fills the yield container (the seeds). Seed filling continues until the seeds reach physiological maturity (seeds reach their maximum dry weight).

Yield is a function of the total rate ($g m^{-2} day^{-1}$) and duration (days) of seed growth during Stage III, thus modifying either of these components will affect yield. The rate of growth is determined by the capacity of the vegetative portion of the crop to supply assimilate to the seeds (photosynthesis and remobilization of stored carbohydrates and N-containing compounds) and the capacity of the seeds to utilize the assimilate to synthesize the storage compounds that give seeds their value. Since seed number is fixed at the beginning of the seed-filling period, any changes in the environment during Stage III will probably affect seed size.

The seed-filling period has two very interesting characteristics, both of which have the potential to influence yield. First, the seed-filling period, when all yield is produced, is relatively short, usually covering roughly 30 to 40 days. The total growing cycle (planting to maturity) of most grain crops is often 100 to 130 days (ignoring winter-grown grains), so only 40% or less of the growth cycle is spent producing yield. Producing high yields in such a short time puts a lot of stress on the vegetative plant to supply the necessary assimilate. Any disruption of the productivity of the crop during this time can have significant effects on yield. Given the shortness of the period and its importance, it is not surprising that increasing the length of the seed-filling period will increase yield. Grain crops do not use time very efficiently when they spend more time on the preliminary activities and less on the actual production of yield.

The second interesting characteristic of the seed-filling period is that shortly after its beginning, the photosynthetic machinery in the leaves starts to destroy itself. Just when the crop is finally producing yield, when the photosynthetic capacity is desperately needed, the leaves start to senesce (as discussed in Chapter 2, this volume), ultimately reducing photosynthesis to zero. The N released by the breakdown of the enzymes in the leaves is translocated to the seeds, as are the carbohydrates stored in leaves and other plant parts. These remobilized materials probably make up for some of the decline in canopy photosynthesis. Senescence is not all bad; it results in very efficient use of N in vegetative tissues instead of discarding it in the stover. It is important to note that not all of the N in vegetative tissues is remobilized. N concentration in mature soybean leaves, petioles and stems was roughly one-half of the concentrations at growth stage R5 (beginning seed fill) (Zeiher et al., 1982). Varieties that have delayed senescence (often called 'stay green' varieties) may have longer seed-filling periods.

The size of the yield container (number of seeds per unit area) is fixed during Stage II and it is filled during Stage III. Ideally, as discussed previously, the number of seeds would be set at a level that matches the capacity of the crop to fill the seeds. Environmental conditions and crop productivity that are the same during both stages would meet this goal, resulting in a normal-sized seed for the variety used. Yield and seed number could be high or low depending upon the productivity of the environment, but seed size would be normal as long as the environment was constant during both stages. Stage II, however, occurs before Stage III and this separation in time could create differences in environmental conditions and crop productivity during the two stages that would upset the balance.

The crop would be unable to fill all the seeds to their normal size if unfavourable conditions and a reduction in growth developed during Stage III, resulting in smaller seeds and lower yield. The reverse situation (unfavourable conditions during Stage II followed by favourable conditions during Stage III) should result in excess capacity to fill the reduced seed number. This excess capacity could create a sink limitation where the capacity of the seeds to grow limits yield. Increases in seed size in this situation could alleviate some of the sink limitation, depending on the capacity of the seed to increase above its normal size. In general, legume seeds probably show more flexibility in this regard than cereals or maize (as discussed in Chapter 2, this volume). The general consistency of seed size among environments suggests that there is usually some persistence in the environmental conditions across stages. Large shifts between Stage II and Stage III are probably more likely on soils with low water-holding capacity where short periods without rain could cause rapidly developing stress. Of course, consistent stress or high productivity during both Stage II and Stage III should result in a reduction or an increase in seed number, but the production of a normal-sized seed. Once again, seed size is not a sure-fire predictor of yield.

Normal seasonal variation in crop productivity also affects the relationship between seed number and the capacity of the crop to fill the seeds. The critical period for the determination of seed number of winter-grown grain crops (e.g. winter wheat) occurs near the spring equinox and thus solar radiation and potential productivity increase from Stage II to Stage III, thereby generally favouring seed filling over the determination of seed number. Summer-grown grain crops are just the opposite: solar radiation is usually higher during the determination of seed number near the summer solstice and then it declines during seed filling, thus favouring seed number. Winter crops (e.g. winter wheat) may be more likely to be sink limited than summer crops (e.g. soybean and maize). Day-to-day variation in solar radiation and other aspects of the environment in many years probably obscures the effect of these average seasonal trends on the balance between seed number and the capacity of the crop to fill the seeds. These differential seasonal trends do not negate the pre-eminent role of seed number in all crops in responding to fluctuations in environmental conditions and potential productivity.

Seed-fill duration is known to be under genetic control in many crops, including soybean, maize, wheat, rice, barley, oat, sorghum, sunflower and common bean (Egli, 2004; Egli, 2017, p. 63) and probably in other grain legumes. Lengthening the seed-fill duration contributed to genetic improvement of yield in many crops. Interestingly, this increase occurred when plant breeders were selecting for yield, not for a longer seed-filling period. A longer seed-filling period would, of course, be associated with a delay in leaf senescence to provide the raw materials for seed growth and a seed that can grow for a longer time.

The length of the seed-filling period is influenced by temperature and moisture stress, which, in turn, affects seed size and yield. Lowering temperature results in a longer seed-filling period that, because yield is related to seed-fill duration, often results in higher yields (see discussion in Chapter 2, this volume). The late W.G. Duncan, a noted crop physiologist at the University of Kentucky, theorized that extra-high yields would occur under irrigation (no water stress) at high elevations in arid environments. The high elevation and arid environment would result in high levels of solar radiation (fewer clouds) and a large diurnal temperature range – warm enough in the day to maximize photosynthesis but cool enough at night to create a long seed-filling period (Duncan *et al.*, 1973). Exceptionally high irrigated maize yields at high elevations in Colorado support this proposition (Muchow *et al.*, 1990).

Drought stress during seed filling accelerates leaf senescence, shortens the seed-filling period and reduces seed size and yield. The accelerated leaf senescence could not be reversed by re-watering soybean plants after only 3 days of stress (Brevedan and Egli, 2003). If only short periods of stress are required to shorten the seed-fill duration and reduce yield, this stress may be a more important yield-limiting factor than is commonly realized. This stress is a 'hidden' stress in the sense that the senescence process appears to be completely normal – it just occurs sooner; without well-watered plants for comparison, the producer is unaware that stress occurred until he harvests the crop and finds smaller seeds and lower yields. Soybean plants growing on eroded soils on knolls often mature early, which is probably a visual manifestation of moisture stress accelerating leaf senescence. Ultra-high grain crop yields may require a complete absence of water stress during seed filling.

Other stresses during seed filling can also shorten the seed-filling period and reduce yield. Wheat responded to waterlogging at jointing and/or anthesis by shortening the seed-fill duration. Nutrient stress (N, P, K) may accelerate senescence and shorten the seed-filling period, as can leaf diseases, although it may be possible to reverse the effect of leaf disease with foliar fungicides. Again, since these stresses simply accelerate the normal senescence processes, they may not be obvious to the producer. The lower yield may come as a complete surprise.

Yield is produced, i.e. the yield container is filled, during the seed-filling period (Murata's Stage III). Consequently, it is an important stage. The old philosophy that yield is 'made' early in seed filling is not necessarily true – the potential is there but stress can easily reduce it. Yield is not 'made' until the seeds approach physiological maturity.

Radiation-Use Efficiency (RUE)

The capacity of a crop to convert solar radiation into plant tissue is often characterized by the radiation-use efficiency (RUE) – the amount of dry matter produced per unit of intercepted solar radiation (units are usually grams of dry matter per megajoule of intercepted radiation, $g\,MJ^{-1}$). Thus, RUE is a measure of how efficiently the plant, via photosynthesis and respiration, uses intercepted solar radiation to produce plant tissues. RUE is useful because it provides a single number that expresses the capacity of the crop community to accumulate dry matter. The single number makes it easier to compare the efficiency of growth among environments and crop species. Its single number characteristic is also its great weakness; the individual characteristics of the processes responsible for crop growth are lost when they are reduced to a single number. The RUE is a measure of efficiency (output per unit input), so it does not determine the rate of dry matter accumulation by the crop community (i.e. CGR); CGR is determined by the RUE, the level of solar radiation and the proportion that is intercepted by the crop community.

Estimating RUE requires measurement of the increase in dry weight of the crop community over time and the amount of intercepted solar radiation. Estimates of RUE are usually based on above-ground dry matter because measuring root dry weight is difficult, time-consuming and notoriously inaccurate. RUE may be calculated using total solar radiation data or photosynthetically active radiation (PAR). PAR is roughly half of total solar radiation, so RUE calculated using PAR will be larger than if total solar radiation is used, which can be confusing if the basis of calculation is not known.

Since RUE is determined, in part, by the rate of community photosynthesis, it is not surprising that crops with C_4 photosynthesis have higher RUEs in favourable environments than crops with C_3 photosynthesis (maximum RUE, based on total solar radiation, averaged $1.54\,g\,MJ^{-1}$ for maize, $0.88\,g\,MJ^{-1}$ for soybean and $1.13\,g\,MJ^{-1}$ for rice) (Sinclair and Muchow, 1999). Community characteristics that influence photosynthesis (e.g. leaf angle) affect RUE. Composition of plant tissues will also affect RUE because of the varying amounts of respiration required to produce oil, protein and complex carbohydrates (see Chapter 2, this volume). Consequently, the RUE of a C_3 cereal (e.g. rice) would be higher than that of a C_3 legume (e.g. soybean) that produces higher levels of oil and protein. RUE will be lower if photosynthesis is reduced by environmental stress (e.g. water or nutrient stress, or leaf diseases).

RUE conveniently provides a single number characterizing crop growth as a function of intercepted solar radiation, making it an attractive concept for crop modellers. Its usefulness is limited, however, by the variability associated with estimates of dry matter accumulation by the crop community. Dry matter accumulation by a crop community is estimated by taking repeated samples of the above-ground dry matter, a labour-intensive measurement that is not very accurate, making it difficult to detect small differences in RUE that could affect yield. Therefore, RUE is rarely used to compare varieties, hybrids or management practices; its use is usually restricted to species comparisons in environments where water and nutrients are not limiting and there are no known stresses limiting plant growth.

Harvest Index (HI)

Harvest index (HI) (Eqn 3.5) is a commonly used indicator of the allocation or partitioning of the products of photosynthesis between vegetative plant parts and yield (seeds):

$$\text{HI} = \text{yield/total biomass at maturity} \tag{3.5}$$

The total biomass includes the seeds and the above-ground vegetative plant parts (leaves, petioles and stems) at maturity. Roots are seldom included in measurements of total biomass. The leaves and petioles that abscise before maturity in some species (e.g. soybean) should be collected and included in the total biomass but are often ignored. We can rewrite the definition of HI as yield/(yield + vegetative plant parts), which illustrates clearly the appearance of yield in both the numerator and denominator of this ratio.

The products of photosynthesis must be partitioned or divided on a daily basis among the various growing organs (leaves, stems, seeds, roots) and the numerous growth processes (respiration, N acquisition, synthesis of starch, protein, oil and storage carbohydrates, etc.). Partitioning changes during crop growth as the plant increases in size and shifts from purely vegetative growth, through a combination of vegetative and reproductive growth, to the final stage where the vegetative plant is senescing and seed growth dominates. Partitioning is complex and dynamic; partitioning patterns probably change diurnally and obviously during crop development. Crop physiologists do not understand the physiological mechanisms regulating partitioning; however, plants manage this complex problem very well. Seldom do they produce too much stem and not enough leaves, too many or too few seeds, or more N than needed. HI simply represents the result, measured at maturity, of the partitioning that occurs throughout the growth of the crop.

The use of HI as an indicator of partitioning was popularized in the 1960s by C.M. Donald (Donald, 1962), an Australian plant breeder/crop physiologist working with wheat. The finding that genetic improvement of yield of wheat, barley and rice was often a result of increases in HI with no change in total biomass (productivity or the capacity to accumulate dry matter did not change) (Austin *et al.*, 1982) further stimulated interest in this ratio. These results suggested that higher yields were simply a result of partitioning more assimilate to reproductive growth instead of increasing the production of assimilate. Suggestions that there might be a maximum HI that could not be exceeded raised the disturbing possibility that yield increases may eventually stop at some time in the future (Austin *et al.*, 1982). It must be noted that there are crops (maize and soybean, for example) that did not rely on changes in HI for their historical yield increases and more recent reports suggest that wheat yields are now increasing with no change in HI (Shearman *et al.*, 2005).

Although the HI is a simple ratio, understanding why it changes is not as simple and straightforward as one might expect. First, changes in this ratio, as in all other ratios, can be misleading and must be interpreted carefully. For example, an increase in HI occurs if yield increases and vegetative weight is constant or declines (the traditional finding with wheat), but an increase can also occur if yield is constant and vegetative weight declines. Such is the nature of ratios – the relationship between vegetative weight and yield changed in both examples but only one of the changes was associated with an increase in yield. Second, HI is measured at maturity, so, like yield, in a sense it is a summary of what happened during the entire growth cycle of the crop. It provides no information about why the index changed, so it does not help us understand the yield production process. Finally, HI of crops whose leaves abscise before maturity (soybean, for example) is often based on an incomplete measure of vegetative weight (often only stem material), which may be misleading. To avoid this problem, the vegetative weight at the end of vegetative growth (growth stage R5 in soybean), representing maximum vegetative weight, is sometimes used to calculate an apparent HI.

HI is usually interpreted as a measure of partitioning of dry matter between vegetative and reproductive plant parts, implying, in the purest sense of the term, that the assimilate produced on any day can be allocated to vegetative or reproductive growth. There are, however, other aspects of crop growth that cause changes in HI without direct changes in partitioning between plant parts:

1. Since vegetative and reproductive growth are largely separated in time, environmental conditions could, as discussed previously, favour one more than the other, resulting in changes in HI. Stress during vegetative growth could reduce vegetative weight without affecting yield, resulting in an increase in HI that, strictly speaking, is not related to partitioning. Conversely, stress during seed filling that reduces yield without affecting vegetative weight will reduce HI, again without a direct effect on partitioning. Management practices that differentially affect vegetative weight and yield (e.g. N fertilization) cause changes in HI.

2. Variation in the length of the crop growth cycle will often affect HI. This variation is often a result of differences in the duration of vegetative growth (Egli, 2004). Maximum vegetative mass of early-maturing varieties is usually less than that of late varieties, but these differences are not necessarily reflected in yield, resulting in changes in HI that are related to the length of the growth cycle but not to yield. The apparent HI of soybean varieties from Maturity Groups II (early) to V (late) decreased as the duration of vegetative growth and vegetative weight increased with no change in yield (Fig. 3.7). HI of sunflower and rice also decreased as growth duration increased (Egli, 2017, p. 149). This apparent change in partitioning was a result of variation in the length of vegetative growth, again not exactly a 'true' change in partitioning.

3. As discussed previously, the length of the seed-filling period is related to yield in many crops. If the duration of seed filling is not related to maximum vegetative weight (there is no compelling reason for such a relationship), higher yield associated with a longer seed-filling period would result in an increase in HI (yield up, vegetative weight constant). Again, this change does not represent a true dividing of assimilate between two potential sinks.

An increase in HI does not necessarily mean that the daily products of photosynthesis were directed to reproductive instead of vegetative growth. Changes can occur that are unrelated to partitioning and that is the curse of HI; finding useful meaning in HI data can be difficult.

In summary, HI is widely used as an indicator of partitioning of assimilate between vegetative and reproductive growth. The usual assumption is that increased partitioning to the seed (higher HI) will increase yield, but HI is an imperfect, complex indicator of partitioning that is not necessarily related to yield. The HI per se tells us little about the plant processes responsible for changes in yield. Perhaps this complexity explains the failure of HI to fulfil its potential, predicted by Donald (1962), as a selection index for higher yield. It is probably more beneficial to deal directly with the processes involved in the production of yield and ignore this simple ratio (Charles-Edwards, 1982, pp. 111–112).

Time and Crop Productivity

Time is an important determinant of crop productivity, but most discussions of productivity focus on rates, such as the CGR ($g\,m^{-2}\,day^{-1}$) or the rate of photosynthesis ($\mu mol\,CO_2\,m^{-2}\,h^{-1}$). The final vegetative weight or yield, however, is always a function of a rate expressed over a finite time. Yield, for example, is determined in Murata's Stage III by the total rate of seed growth ($g\,m^{-2}\,day^{-1}$) and the number of days that seed growth continues (seed-fill duration). Both rate and time are important determinants of yield. The time available for crop growth (often taken as the length of the frost-free season in temperate climates)

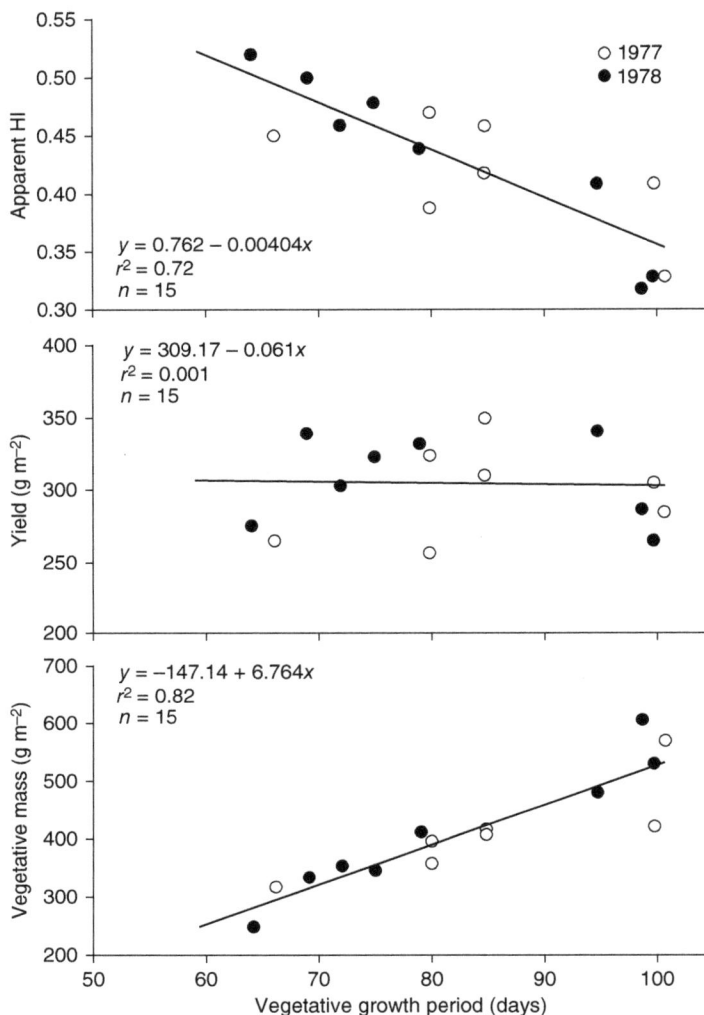

Fig. 3.7. The relationship between length of the vegetative growth period (planting to growth stage R5) and apparent harvest index (HI), yield and maximum vegetative weight (determined at growth stage R5, the beginning of seed fill). Eight soybean varieties from Maturity Groups II to V were grown in the field for 2 years. The apparent HI index is the ratio of yield to maximum vegetative weight + yield. (Adapted from Zeiher et al., 1982.)

is also important in determining where crops are grown and what cropping systems can be utilized at a given location. The rising temperatures associated with climate change are making more time available for crop growth, which will open up new areas for grain production at higher latitudes (Linderholm, 2006).

The time available for crop growth, usually set by temperature or water availability, determines the total solar radiation available for crop growth and thus the potential productivity. Potential productivity is defined as the accumulated solar radiation available during the time when temperatures are suitable for plant growth. The combination of time and solar radiation determines the maximum productivity possible at any location. Crop plants grow only when temperatures are in the appropriate range and their growth during this period is ultimately limited by solar radiation. Potential productivity is defined by two aspects of the environment that cannot be modified by producers. Producers routinely manipulate, for example, soil fertility and often water availability, but temperature and solar radiation at a location are beyond their control. The availability of water, instead of temperature, may define the time component of potential productivity for rain-fed agriculture in tropical climates with distinct wet and dry seasons.

Potential productivity defines the maximum productivity possible, but actual productivity is usually less than the potential due to limitations of the crop or the environment. All sports fans are acquainted with players who do not perform up to their potential; crops or cropping systems are no different, their productivity often does not reach the potential set by time and solar radiation. The concept of potential productivity is useful to compare locations and to think about what crop or cropping system would utilize more of the potential productivity.

Time is an important determinant of potential productivity. Potential productivity (summation of the average solar radiation during the average frost-free season) more than doubled from north to south across the maize belt in the Midwestern USA (~90°W longitude) (from the Canadian border, approximately 49°N latitude, to New Orleans, Louisiana, 29°N) (Fig. 3.8). Most of this increase was a result of an increase in length of the growing season, i.e. time. By comparison, potential productivity in the Cerrado region of Brazil at 14°S latitude with a 365-day growing season was 6900 MJ m^{-2}. Potential productivity would probably decrease as elevation above sea level increased (shorter growing season, but perhaps more solar radiation) and it is increasing as the rising temperatures associated with climate change create longer growing seasons. There are obviously large variations in potential productivity among locations; the challenge is to effectively convert this potential into actual productivity with grain crops.

The total growth duration of cultivated grain crops exhibits tremendous variation, matching, in a sense, the variation in potential productivity. There are varieties with durations as short as 62 days (cowpea) and as long as 185 days (sorghum) (Egli, 2011). A sorghum landrace from Ethiopia matured after 240 days and a common bean variety required 200 days to mature. There is also substantial variation among varieties within a species; for example, soybean varieties varied from 86 to 141 days and maize hybrids from 78 to 149 days (Egli, 2011). A commercial company recently released a 69-day maize hybrid, further widening the range in maize (Anonymous, 2017). Surprisingly, this

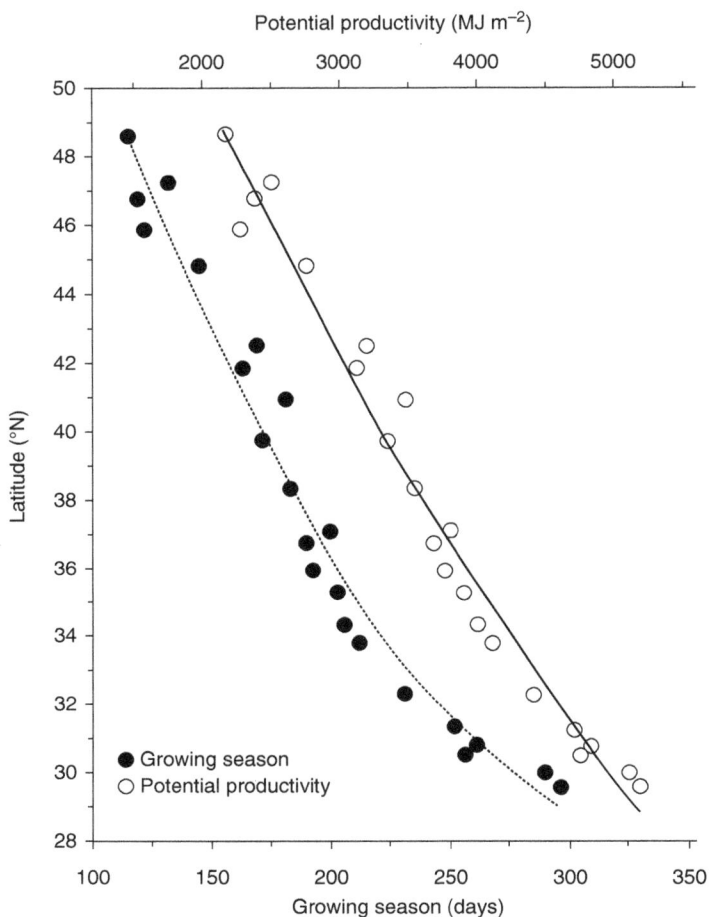

Fig. 3.8. Variation in time and potential productivity from north to south across the maize–soybean belt in the central USA (~90°W longitude). The growing season is the average number of days (1971–2000) from the last frost (0°C (32°F)) in the spring to the first frost in the autumn (NOAA, 2016). Potential productivity is the summation of average (1998–2008) solar radiation (MJ m⁻²) during the growing season (CRA, 2016). (From Egli, 2017, p. 152.)

variation in total duration does not necessarily relate to grain yield. Selecting the variety that has the longest growth duration does not guarantee high yield.

The relationship between total growth duration and the length of vegetative (Stage I) and reproductive growth stages (Stages II and III) is the key to understanding how time affects productivity and yield. The length of the vegetative growth phase increases in step with the total growth duration (Fig. 3.9a) for a large number of grain crops. The maximum vegetative weight increases as the duration of vegetative growth increases (Fig. 3.10), because late-maturing varieties have more time to produce vegetative tissues than early varieties. If yield is the total above-ground weight at maturity (total biomass), the longer crops grow, the higher the yield (Fig. 3.10). A crop that matures after 90 days of growth will not produce nearly as much biomass as a crop that grows for 120 or 365 days in the tropics. Unfortunately, the relationship between grain yield and time is much more complicated.

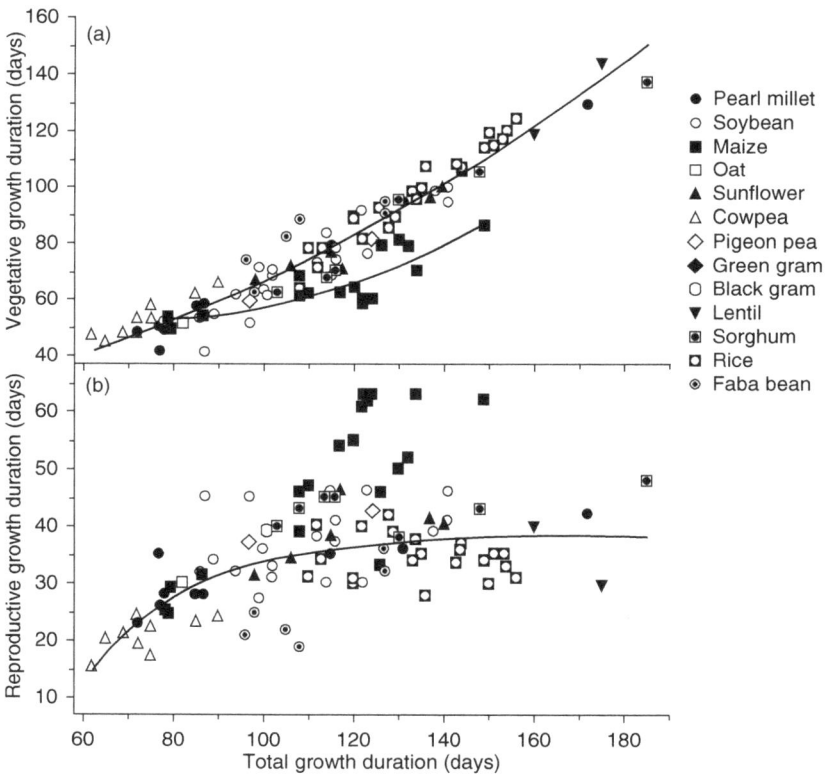

Fig. 3.9. The relationship between total growth duration (days from planting or emergence to maturity) and (a) duration of vegetative growth (days from planting or emergence to flowering or the beginning of seed filling) and (b) duration of reproductive growth (flowering or the beginning of seed filling to maturity) for 13 grain crop species. The original data sources can be found in Egli (2011). Regression models: for all species except maize in (a), $y = 18.68 + 0.21x + 0.0027x^2$, $r^2 = 0.94$ ($P < 0.001$), $n = 86$; for maize in (a), $y = 83.58 - 0.85x + 0.006x^2$, $r^2 = 0.71$ ($P < 0.001$), $n = 18$; for all species except maize in (b), $y = -306.52 + 344.83(1 - e^{-0.0432x})$, $r^2 = 0.43$ ($P < 0.001$), $n = 86$. (Adapted from Egli, 2011. Used with permission from John Wiley and Sons.)

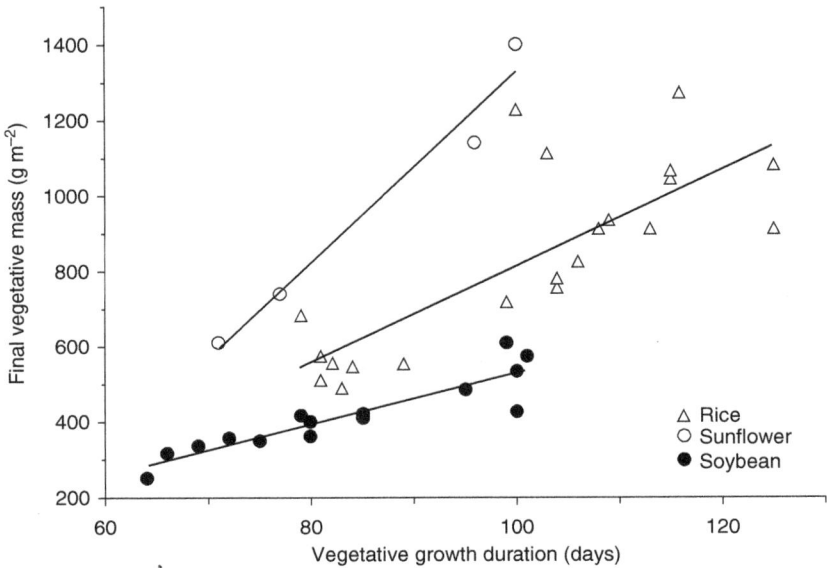

Fig. 3.10. The relationship between maximum vegetative weight at the beginning of seed fill (soybean) or final vegetative weight (sunflower and rice) and the duration of the vegetative growth period (days from planting to flowering or beginning seed filling). Regression models and data source: soybean, $y = -152.32 + 6.83x$, $r^2 = 0.82$ ($P < 0.001$), $n = 15$ (Zeiher *et al.*, 1982); sunflower, $y = -1197.50 + 25.23x$, $r^2 = 0.97$ ($P < 0.05$), $n = 4$ (Villalobos *et al.*, 1994); rice, $y = -466.29 + 12.80x$, $r^2 = 0.62$ ($P < 0.001$), $n = 22$ (Venkateswarlu *et al.*, 1977). (From Egli, 2011. Used with permission from John Wiley and Sons.)

The duration of reproductive growth did not track the total growth duration; it reached a maximum at a total growth duration of 100 to 120 days with no further change as the total growth duration increased to nearly 180 days (Fig. 3.9b). With no change in the duration of reproductive growth, there is no reason to expect higher yield as growth duration increases beyond 100 to 120 days, unless the longer growth duration positions reproductive growth in a more favourable environment – an unlikely occurrence. In fact, longer total growth durations may position reproductive growth in a less favourable environment (lower solar radiation and lower temperatures). Longer total growth durations will increase maximum vegetative weight, but that will not necessarily increase yield. Yield is limited by the duration of reproductive growth, so using varieties with longer total growth cycles in environments with large potential productivity would probably not increase yield.

The failure of the seed-filling period to lengthen in step with the total growth duration greatly restricts the capacity of grain crops to exploit the environment once the length of the total growth duration exceeds 100 to 120 days. The length of the growing season and potential productivity

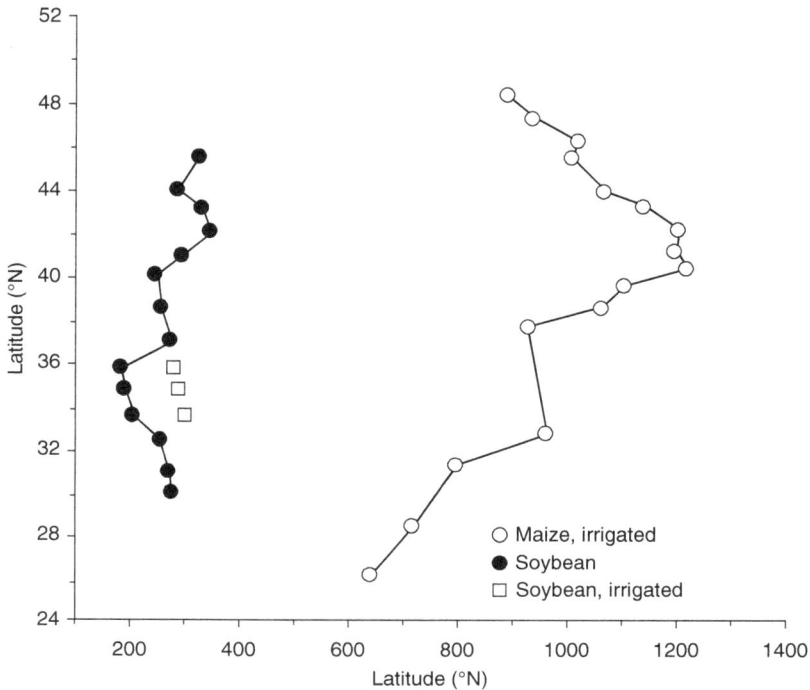

Fig. 3.11. Average yields of maize and soybean (2005–2014) on a transect from north to south across the central USA. Soybean data are averages by selected crop-reporting districts from Minnesota, Iowa, Missouri, Arkansas (irrigated data from Arkansas) and Louisiana. Irrigated maize data are from North and South Dakota, Nebraska, Oklahoma and North Texas. All data from NASS (2020).

increase as we move south across the central USA (Fig. 3.8) but yields of maize and soybean actually decrease along the same transect (Fig. 3.11); varieties with longer growth durations are used at lower latitudes, but they cannot convert the higher potential productivity into more yield. This pattern was maintained when irrigation minimized the effects of water stress on yield. The failure of grain crops to utilize the higher potential productivity at lower latitudes is not entirely a result of limitations from reproductive growth duration – poorer-quality soils and greater prevalence of diseases and insects are probably also involved. The old axiom to use varieties that utilize most of the growing season is not always valid. The relatively inflexible nature of the length of reproductive growth as the growth cycle increases results in a very inefficient use of time by grain crops; they often spend too much time producing unneeded vegetative mass instead of producing yield. This inefficiency has many important implications for crop productivity and management.

The inability of a single crop to fully utilize the potential productivity at locations with long growing seasons (i.e. high potential productivity) can be overcome by growing more than one crop in a year (multiple cropping). Multiple cropping utilizes more of the potential productivity by essentially creating several seed-filling periods in a single year, overcoming the problem of the inflexible seed-fill duration. Multiple cropping is widely used in temperate (e.g. growing soybean after winter wheat in the mid-south of the USA) and tropical climates with 365-day growing seasons (e.g. rice–wheat systems or multiple rice crops).

The development of short-duration varieties was a key to successful multiple-cropping systems involving rice, mung bean and cowpea (Egli, 2017, p. 159). Shortening the vegetative growth period without shortening the reproductive growth period (Fig. 3.9) makes it possible to have the best of both worlds – short duration for multiple cropping without sacrificing yield potential. Ratooning (managing regrowth from a harvested crop to produce a second crop) for grain or fodder of, for example, sorghum, pearl millet or rice accomplishes the same objective as a multiple-cropping system. Interestingly, the rising temperatures associated with climate change should lengthen growing seasons and expand the opportunities for multiple cropping.

Although the total productivity of a multiple-cropping system is higher than a single-crop system, it may not be as attractive from an economic standpoint. The production costs (seed, planting, weed and pest control, harvesting) of each crop do not change much, but the yield of the second crop in the sequence is often reduced below its yield when grown as a single crop. Static costs and lower yields may reduce profits and the attractiveness of the system to the producer.

The relationship shown in Fig. 3.9 has a number of important implications for crop management beyond the utilization of time just discussed. It is obvious from the relationship in Fig. 3.9 that the length of the total growth cycle can be reduced up to a point without necessarily reducing yield. Reducing the total growth cycle from, for example, 140 to 120 days would have no effect on the length of reproductive growth and yield, but it would shorten the vegetative growth period and reduce the maximum vegetative weight and LAI. Shortening the total growth cycle would reduce the total water use, an important benefit in irrigated production systems that would contribute to a more sustainable system as water scarcities increase. A shorter growth cycle also increases the opportunity to manipulate planting dates to position crop growth in the most favourable environment, thereby avoiding environmental stress. For example, the very successful Early Soybean Production System (ESPS) dramatically increased yield in mid-south regions of the USA by using early-maturing varieties planted early to avoid late-season water stress (Heatherly, 1999). Short-duration varieties can, in subsistence production systems, provide an important source of food early in the current cropping season when stored food supplies are depleted. Reducing the duration of vegetative growth, however, may make the crop more susceptible to stress. A smaller vegetative plant will increase the likelihood that stress could reduce the LAI below the critical level, reducing solar radiation interception, CGR and yield.

There are, of course, some advantages associated with using varieties with long vegetative growth periods, even though they do not directly increase yield (Egli, 2017, pp. 159–160). The larger vegetative plants associated with long vegetative growth periods may have been needed for maximum solar radiation interception in cropping systems with relatively wide row spacings. Wide rows (1.0 m (40 in)) were originally needed so horses could walk between the rows and for mechanical weed control. These traditional practices may have been the source of recommendations to use full-season varieties. Mechanical cultivation is not needed in modern cropping systems, so crops are grown in narrower rows and maximum solar radiation interception can be achieved with earlier varieties that have shorter vegetative growth periods and lower maximum LAIs. Modern narrow-row cultural systems have virtually eliminated the need for the large vegetative weights and high LAIs produced by long-duration varieties.

Long vegetative growth periods would be advantageous in production systems that harvest both grain and stover. Since root growth is usually associated with vegetative growth, i.e. roots stop growing at the end of Stage I, long vegetative growth periods should result in more and possibly deeper root systems, providing expanded access to water and nutrients. Large vegetative plants store more N and carbohydrates that can be translocated to the seeds during seed filling, which may help mitigate stress-induced yield loss. Long vegetative growth periods may have a negative effect in water-limited environments by exhausting water supplies during vegetative growth, leaving little for use during grain filling.

There are many situations where either long- or short-duration varieties provide a yield advantage or a more efficient production system. These advantages are specific for species, environments and cultural systems, but the use of short-duration varieties to enable multiple-cropping systems or to position reproductive growth in the most favourable environment may be two of the most important, while reducing water use may become more important in the water-limited environments of the future. The potential value of long- or short-duration varieties should not be allowed to obscure the basic principle that there is no inherent relationship between yield and total growth duration once it exceeds the minimum needed for maximum yield.

Summary

Our goal in this chapter was to develop a general model of community growth and the production of yield by grain crops. Murata's (1969) three-stage system provides such a model. It is useful because it is simple (only three stages), it applies equally well to all grain crop species (although there is some species variation in minor details), it clearly identifies the sequential nature of the yield production process and the three stages relate to the primary drivers of the yield production process at the community level. First, the crop must accumulate the leaf area that drives community photosynthesis (Stage I), then seed number is determined (Stage II), and finally seed filling occurs (Stage III) and the production of yield is finished. High yield of any variety/location combination requires, at a minimum: (i) the production of enough LAI during Stage I to maximize solar radiation interception and community photosynthesis; and (ii) an absence of stress during Stage II to maximize seed number and during Stage III to allow the seeds to fill to their maximum potential size.

The scheme provides a powerful framework for us to think about how management decisions and environmental conditions affect yield. A framework or, if you will, a model is useful because we can better anticipate the results of our actions on yield. Simply focusing on yield, as in 'do narrow rows increase soybean yield?', usually requires extensive time-consuming experimentation to get the answer that is no help to the producer who must make an immediate decision. Murata's three stage model will not provide a direct answer, but it helps the producer identify the critical aspects of the question that may lead to more informed decisions (if wide rows provide complete ground cover before the beginning of reproductive growth (Stage II), it is unlikely that there will be a response to narrowing the rows). At the very least, it will guide research efforts by helping ensure that the treatments chosen and the data collected (solar radiation interception at the beginning of reproductive growth in this example) will provide an answer at the mechanistic level in the least amount of time. We will rely extensively on Murata's model in the next chapter as we consider fundamental principles underlying the effect of common management practices on yield.

Crop Management: Principles and Practices

4

Live as if you would die tomorrow,
Farm as if you would live forever.
Old Farmer's Proverb

Introduction

Successful grain crop production hinges on management. Planting date, variety, row spacing and plant population are among the choices made annually by all grain crop producers. Other aspects of the production system, including location (climate and soil type), crop species, use of irrigation, choice of equipment and tillage system, are less flexible in the short term and often don't change from year to year. Many, if not all, choices are affected by external factors beyond the manager's control, including variation in weather conditions, economics (cost of land and inputs, price of grain), government programmes, trade issues, consumer preferences and competition from production in other countries.

Management decisions are always made in this unpredictable, constantly changing, matrix of factors that affect inputs, yield and profits. Some management decisions must be made when many aspects of the system are unknown; for example, weather conditions during the growing season, potential yield or grain prices at harvest. Consequently, decisions are often based on the expectation of typical or average conditions (or better or worse depending upon whether the producer is an optimist or a pessimist) during the growing season. Such decisions may prove to be 'wrong' as the growing season unfolds, leading to lost yield or inefficient use of resources. Big data, algorithms, artificial intelligence, crop simulation models and accurate long-range forecasts of growing-season weather may eventually reduce this uncertainty and make the producer's job much easier.

Grain crop production is simpler and easier to manage than many high-value crops, principally because fewer time-sensitive decisions are required when the crop is growing. Exceptions to this statement include irrigation

© D.B. Egli 2021. *Applied Crop Physiology: Understanding the Fundamentals of Grain Crop Management* (D.B. Egli)
DOI: 10.1079/9781789245950.0004

timing in irrigated systems, weed, disease and insect management and in-season fertilizer applications for some crops. In this chapter we will focus on decisions made before the crop is planted, so we will discuss planting-seed quality, variety selection, plant population, planting date and row spacing. We will not discuss soil fertility and fertilizer applications, or weed, disease and insect control. These are very important aspects of any management programme, but they are not closely related to crop physiology, so they are beyond the scope of this book and certainly beyond the expertise of the author.

The fundamental goal of crop management is to improve the above- and below-ground environment that the crop is grown in; or, to put it another way, to remove negative aspects of the environment that reduce crop growth and yield. When producers control weeds, adjust planting date or irrigate to reduce water stress they are improving the environment the crop is growing in, i.e. the crop is no longer limited by weed competition, an unfavourable environment created by an incorrect planting date or a lack of water. In a sense, the producer's goal is to create the perfect environment for the growth of the crop where the only limiting factors are the capacity of the plant to convert solar radiation into plant tissue and aspects of the environment (solar radiation levels, CO_2 concentration and temperature) that influence crop growth but are beyond the direct control of the producer.

Yield of a crop in this 'perfect' environment is often considered to be the potential yield, i.e. 'the yield of a cultivar (variety) when grown in environments to which it is adapted; with water and nutrients not limiting; and with pests, diseases, weeds, lodging and other stresses effectively controlled' (Evans, 1993, p. 292). Potential yield is only limited by the plant and solar radiation, CO_2 concentration and temperature. The yield gap (potential yield – producer yield) is a measure of how close yields in the producer's fields are to the potential. It is very unlikely (especially if one considers the producer's desire to make a profit) that management can produce the 'perfect' environment and reduce the yield gap to zero. Some (Fischer *et al.*, 2014, p. 33) argue that increasing yield above roughly 80% of the potential yield is practically difficult and not economical.

Yield does not respond continuously to most management practices. Yield increases in response to the application of the practice, but when the negative aspect of the environment is eliminated, there will be no additional response. Once the appropriate row spacing is deployed, for example, no further progress is possible or once water stress is eliminated, no further yield gains can be obtained by irrigation. This saturation-type response makes it difficult for management to continue to increase yield year after year, because the increase requires a continuous removal of some negative aspect of the environment. Management practices that attack new negative aspects of the environment are needed to drive a perpetual increase in yield. Variety selection, a critical component of any management system, is an exception to this saturation-type response because varieties experience continuous improvement. The yield of today's varieties is higher than they were 10 years ago, and they will be even higher 10 years into the future. Variety selection does not fit very well into the concept of eliminating the negative aspects of the environment, except when a new variety with disease or insect resistance produces higher yield.

Some scientists argue that our major grain crops and their highly productive cropping systems are approaching the perfect environment, which may limit future yield increases. While one might think that there must be a limit to the number of negative factors in any environment, a changing climate and the appearance of new diseases, troublesome weed species and insects will maintain a ready supply of challenges for crop management. Development of new technology may also make it possible to deal with previously untouchable negative factors. Even if, unlikely as it seems, the perfect environment is achieved, varietal improvement will continue to drive yields upward.

The goals of crop management are to produce maximum yield (i.e. to approach the perfect environment), do it efficiently (determined by evaluating output as a function of inputs) and in a manner that minimizes environmental damage. Yield and efficiency are both important to the producer's bottom line, although the latter does not necessarily contribute to higher yields. Finding new management practices that increase yield is becoming increasingly more difficult, but there are probably still many opportunities to improve efficiency. For example, reducing soybean plant population without reducing yield or controlling weeds with fewer herbicide applications while maintaining yield will increase efficiency (fewer inputs, the same yield).

Efficiency and yield are not completely independent. Adopting a management practice that improves efficiency could also increase yield if it improves the quality of the practice. Roundup Ready herbicide technology made weed control less complicated (no need to match herbicides to weed species in each field, no variation in the method of application) and required fewer inputs to achieve satisfactory weed control (was more efficient), which probably resulted in better overall weed control (at least until resistant weeds appeared) and therefore higher yields. Improving efficiency resulted in better outcomes and higher yield.

Grain crop producers often seem to place a higher priority on increasing yield than improving efficiency. Products or practices touted to increase yield are often enthusiastically adopted with only anecdotal evidence of yield improvement. Producers seem to maintain an undying interest in adopting management practices associated with record yields in spite of their uneconomic inefficiencies. Why this philosophy persists, especially when facing low grain prices, is difficult to understand, given that the most important measure of the success of a cropping system is its profitability. Mounting concerns on damage to the environment and interest in sustainability may ultimately force producers to think more about efficiency (i.e. limiting inputs). Both aspects of crop management will come into play as producers struggle to adapt their production systems to new and possibly more extreme environments created by climate change.

The physiological principles describing crop growth and the production of yield, discussed previously in Chapters 2 and 3, provide the basis for informed crop management decisions. All of the primary management practices used in grain crop production are, in some form or fashion, based on these basic processes and physiological concepts. My goal in this chapter is to investigate the physiological foundations of five management practices common to all grain production systems. These physiological foundations apply across all grain crop species and provide a basis for making management decisions that are better informed and more likely to increase yield and/or efficiency, especially when facing new untested management practices. Understanding these physiological principles provides a framework for evaluating proposed practices and discarding those that will likely provide little benefit. This approach could speed up the adoption of new useful technologies and make it possible to reject 'miracle' products or practices without extensive testing. Finally, this chapter will be heavily oriented towards maize and soybean because they are important grain crops and the crops that I know best, although, again, the general principles apply to all grain crops – it is only the details that vary.

Planting-Seed Quality

The American Seed Trade Association's slogan 'first the seed' is a very apt description of the importance of seed in crop production systems. Not only is the seed harvested for yield, it is also planted to regenerate the crop. The quality attributes of planting seed are entirely separate and different from those of seed harvested for yield. Without high-quality planting seed – seed that germinates and produces a seedling that emerges from the soil in a wide range of soil conditions – there will be no crop. Ideally, the seedlings will not just emerge, they will emerge rapidly and uniformly.

The seedbed where germination and emergence occur can be a very hostile environment. Low temperatures, too much or too little water, presence of diseases and insects, poor seed placement (too shallow or too deep), poor seed–soil contact and the development of a crust on the soil surface after planting all have the potential to reduce seedling emergence. No-till and minimum-till cropping systems, cover crops and early plantings may enhance the hostile environment and reduce emergence. We can add poor seed quality to this list. High-quality planting seed is an essential component of seedling emergence.

The germination test is the basic measure of planting-seed quality. The official definition of germination is 'emergence and development from the seed embryo of those essential structures, which, for the kind of seed in question, are indicative of the ability to produce a normal plant under favourable conditions' (AOSA, 2019). Official germination tests follow rules developed by the Association of Official Seed Analysts (AOSA, 2019), with the temperature and the time until the final count of germinated seeds set to maximize germination of the species tested. At the end of the germination test, the seed analyst classifies the seeds/seedlings as dead, abnormal or normal. The proportion of the seeds falling into the normal category is the germination percentage listed on the seed tag of all seed offered for sale.

Since the standard germination test is conducted in favourable conditions for the species in question, germination is usually a good predictor of emergence when the conditions in the seedbed are favourable. If seed with 90% germination is planted at the optimum depth in warm, moist soil and there are no negative factors present (e.g. crusting, disease, insects, etc.), emergence will be roughly 90%. Unfortunately, conditions in the seedbed at planting may not always be favourable and emergence can be significantly less than the germination percentage on the seed tag. The recent trend for early planting of maize and soybean in the maize belt of the USA, along with greater use of no-till and minimum-till cropping systems and cover crops, increases the chances that the seedbed may be cool and wet (i.e. unfavourable), delaying and potentially reducing germination and emergence. Coating seeds with fungicides and/or insecticides before planting, a common practice for both maize and soybean, helps ensure adequate emergence in these unfavourable conditions. These treatments provide relatively cheap insurance against stand failure, although there are concerns with how to handle seed that is treated but not planted.

The concept of seed vigour was developed many years ago as a response, in part, to the failure of seed with high germination percentages to produce adequate stands in less than desirable seedbed conditions. The official definition of seed vigour – 'those seed properties which determine the potential for rapid, uniform emergence and development of normal seedlings under a wide range of field conditions' (AOSA, 2009) – specifically relates to performance in unfavourable seedbeds. The seed vigour concept expands the predictability of the quality test well beyond the 'favourable' conditions targeted by the germination test. Estimates of seed vigour supplement estimates of germination by providing additional information on the potential performance of the seed.

Seed scientists have developed many seed vigour tests since the introduction of this concept, tests that are often associated with specific crop species. Many seed vigour tests mimic less than favourable seedbed conditions by measuring germination of the seed after they are exposed to stress. The accelerated-ageing test (AOSA, 2009), often used on soybean seed, stresses seed in a high-humidity environment at a temperature of 41°C (105.8°F) for 72 h (for soybean) after which germination is determined in the standard germination test. High-vigour seed does not deteriorate during the stress portion of the test and the germination after ageing will be essentially equal to the standard germination (seed lot 1, Table 4.1). As seed vigour decreases, the seed is progressively unable to withstand the stress of the accelerated-ageing test and the difference between accelerated-ageing germination and standard germination increases (Table 4.1). The capacity to withstand the stress provides an index of seed vigour and predicts how the seed will respond when planted in a stressful seedbed.

The cold test was originally developed to estimate seed vigour of maize, although more recently it is also used on soybean. As the name implies, the

Table 4.1. Performance of five soybean seed lots of the same variety planted in the field under stressful conditions. (Adapted from Fabrizius, 1993.)

Seed lot	Standard germination (%)	Accelerated-ageing germination (%)	Seedling emergence (%)[a]
1	98	95	87
2	95	83	72
3	96	66	51
4	99	21	32
5	92	4	12

[a]Seeds were treated with a fungicide before planting.

stress is provided by low temperatures and often the presence of pathogens. There are several variants of this test, but for maize the seeds are always placed on a moist substrate and held at 10°C (50°F) for 7 days. After this stress period, the seeds and substrate are moved to 25°C (77°F) for 5 days and the number of germinated seedlings counted. One variant of this test covers the substrate with soil from a maize field to introduce pathogens into the system, while another variant places the maize seed embryo down on the saturated medium. As in the accelerated-ageing test, high-vigour seeds are better able to withstand the cold (and pathogen stress, if present) and have higher germination after the stress.

Many other tests are available to provide an indication of seed vigour, including some that do not involve stressing the seed. Examples of other tests include an early count on the standard germination test (high-vigour seeds germinate faster), the electrical conductivity test (low-vigour seeds leak more electrolytes into the soak solution resulting in a higher electrical conductivity) and various types of biochemical tests (Black *et al.*, 2006, pp. 741–746).

Standard germination and seed vigour can vary widely among seed lots of the same variety (Table 4.1), suggesting that both characteristics of seed quality are influenced by environmental conditions during production, processing and storage of the seed before planting. What causes this variation in seed quality? Standard germination and seed vigour decline as seeds age (just as humans lose their vigour as they age) and seed vigour declines faster than standard germination. The ageing process starts after the seed reaches physiological maturity on the plant and continues during storage, first on the plant between physiological maturity and harvest (remember, physiological maturity occurs at seed moisture levels well above those where seeds are normally harvested) and then during processing and storage until the seed is planted. The rate of decline in seed quality varies among species and is affected by temperature and seed moisture levels.

Seed that reaches physiological maturity under cool, dry conditions will usually have higher quality than seed that matures when it is hot and wet. Late-maturing soybean varieties, for example, are more likely to reach physiological maturity when it is cool and dry and often have higher seed quality than early varieties. Prompt harvesting of seed as soon as it reaches a harvestable moisture level minimizes quality loss while the seed is in storage on the plant. There are some reports of genetic (variety) differences in these quality characteristics, but they are less likely and smaller than environmental effects. Variety selection cannot be used to reliably find high-quality planting seed. Only the seed tag has that information.

Supply and demand for seed of specific varieties may result in storage of unsold seed for sale the next year. Quality of this carryover seed will continue to decline during storage, although the rate of decline will depend upon the initial quality (standard germination and vigour), species and storage conditions. The decline will be faster in warm conditions (ordinary warehouse in relatively warm climates) and slower at lower temperatures (controlled-climate storage). Seed moisture levels also play an important role in the rate of decline: the drier the seed, the slower the decline in quality. Regardless of the storage environment, a vigour test before planting will identify those seed lots that have deteriorated to the point that they are unlikely to produce adequate stands in the field.

Table 4.1 illustrates the essence of seed vigour with five soybean seed lots of the same variety that all have high standard germination but exhibit a wide range in seed vigour levels (accelerated-ageing germination ranging from 95 to 4%). When these seed lots were planted early in the spring in a stressful seedbed environment, seedling emergence closely tracked the accelerated-ageing germination. Seed lot 4 illustrates the value of seed vigour testing; it appeared to have exceptional quality (standard germination of 99%), but emergence in the field was only 32%, resulting in an unacceptable stand. The accelerated-ageing test identified the potentially poor performance of this seed lot when planted in unfavourable conditions. In favourable field conditions, emergence would have been close to 99%.

It must be noted that the use of high-vigour seed does not guarantee adequate stands. Stress in the seedbed can be so severe that even high-vigour seed will have reduced emergence and possible stand failure. High-vigour seed increases the probability of achieving adequate emergence, but it is never a sure thing.

Does planting high-vigour seed result in higher yields? It might seem logical from the dictionary definition of vigour ('active force or strength; vitality; energy') that plants from high-vigour seed would produce higher yield. The answer, however, is a straightforward yes and no. If the use of high-vigour seed prevents poor emergence and less than adequate stands, yield will be increased above the level that would have been harvested from the poor stand. It would also save the cost and the potential yield loss associated with replanting. If, on the other hand, high-vigour seed does not affect emergence (seed planted in favourable soil conditions where emergence is similar to standard germination), seed vigour will have no effect on yield. Planting high-vigour seed helps ensure against poor stands and the associated losses, but it does not, by itself, increase yield.

Seedlings from high-vigour seed will usually emerge sooner, so they will always be larger than those from low-vigour seed. This advantage, which is primarily a result of earlier emergence, not faster growth, could result in higher yield of species whose yield is the weight of the vegetative plant (TeKrony and Egli, 1991). The yield of grain crops is not related to vegetative plant size (as discussed in Chapter 3, this volume), so the larger plant from earlier emergence provides no benefit.

The time of emergence of individual seedlings from high-vigour seed lots will be more uniform than from low-vigour lots. Variation in the time of emergence within the stand can reduce the yield of maize and other non-plastic species, but it will have no effect on plastic species like soybean. Interestingly, lower seedbed temperatures that delay emergence increase the variation among individual seedling emergence in both maize and soybean (Egli *et al.*, 2010; Egli and Rucker, 2012). Early planting in cool soils could increase non-uniformity and potentially reduce maize yield.

In summary, high-quality planting seed provides the foundation of any successful crop production system. Modern cropping systems that include the potential stress of earlier planting, greater use of minimum- or no-till systems and cover crops, as well as plant populations nearer the minimum required for maximum yield need high-quality planting seed that will produce seedlings that emerge rapidly at high levels in a wide range of seedbed conditions.

Variety Selection

What variety will produce the highest yield? Every grain producer must answer this question and it is an important question, because improved varieties provide the foundation for productive cropping systems. To simplify this discussion, the term 'variety' is used to refer to varieties and hybrids. An improved variety provides higher yield potential, desirable agronomic traits (standability, reduced shattering, etc.) and protection against some diseases and insect infestations. Modern varieties of some crops offer, via genetic engineering, herbicide tolerance traits that simplify weed control (at least before the development of weeds that tolerate those herbicides) and improved insect resistance. Producers have many choices when selecting a variety or hybrid. For example, in 2019 the University of Illinois evaluated 234 soybean varieties and 190 maize hybrids in their variety-testing programme, while the University of Kentucky evaluated 161 soybean varieties and 155 maize hybrids. Varieties and hybrids are constantly being improved, so producers must change their variety or hybrid regularly, selecting from the many available, to take advantage of these improvements.

For millennia, farmers improved their crops by saving seed from the best-looking plants in their fields to plant the next crop. State extension personnel in the pre-hybrid era trained Midwestern maize belt farmers to select the 'best' maize ear to save to plant the next crop (Wallace and Brown, 1988). By the turn of the 20th century, our understanding of plant reproductive biology improved and hybridization between two parents started to be used to create new varieties. Soybean varieties created by hybridization began to replace selections from plant introductions in the 1940s (Hartwig, 1973). Maize hybrids began to replace open-pollinated varieties in the early 1930s and hybrids occupied 90% of the area in the Midwestern maize belt by 1950 (Russell, 1991). These changes initiated the high-input era of agriculture, characterized by a steady increase in yield (Fig. 1.1, Chapter 1, this volume) and greater use of inputs from outside the farming system. This transformation occurred more or less at mid-century in grain production systems in many parts of the world.

Most of the plant breeders working with maize and soybean in the USA at the beginning of the high-input era were employed by land-grant universities or the USDA. The varieties they produced were made available without royalties or fees to producers who were free to save seed from their production fields for planting the next crop. This system changed over time to include greater involvement of commercial plant breeding companies and a correspondingly smaller contribution from land-grant universities and the USDA. The shift in the USA was driven by the widespread adoption of hybrid maize and then the passage of the Plant Variety Protection Act by the federal government in 1969. The use of hybrids protected the proprietary interests of the originating company. Producers could not save seed for the next year's crop and, since the company controlled the parents of the hybrid, there could not be any unauthorized production of the hybrid. The Plant Variety Protection Act assigned proprietary control to the developer of varieties, providing, by legal fiat, the same protection to the developer of non-hybrid varieties. The use of molecular biology techniques to create herbicide- and insect-tolerant varieties (GMOs) by commercial plant breeding companies further accelerated the decline in public variety development activities, so today most of the major grain crop varieties available to producers come from commercial plant breeding companies.

Plant breeders developing improved varieties have two general objectives (at least before the GMO era). One important objective was 'defect elimination' – eliminating plant characteristics ('defects') that were reducing or limiting yield. Plants that did not lodge or shatter their seeds before harvest or had disease or insect resistance produced higher yields than plants that lodged, shattered or were susceptible to disease or insects. Defect elimination increased yield in the producer's field, but it only 'recovered' the yield lost because of the defect. A variety with lodging resistance would not produce higher yield in an environment where there was no lodging; disease resistance was of no value in the absence of the disease. The improvement from defect elimination was limited by the availability of defects: once a defect was eliminated, no further gain could be made unless other defects were available; consequently, defect elimination becomes progressively less useful as variety improvement proceeds. One can draw a direct analogy with the management goal of improving the crop's environment. Increasing yield becomes progressively more difficult in both systems as the availability of defects or the limiting aspects of the environment decrease. Of course, new diseases or insects or a breakdown of resistance often provides a continual supply of defects, just as a changing climate will provide new opportunities for management.

A second approach focuses on improving the inherent ability of the plant to convert solar radiation into plant tissue, thereby increasing the yield potential. The effect of breeding for yield per se is almost the direct opposite of defect elimination: instead of recovering 'lost yield', productivity itself is increased. These two approaches are separate in theory; in practice, the plant breeder often simultaneously selects for yield and against any defects that need eliminating. The results of successful defect elimination are often readily apparent in the field; the producer, for example, can easily see that a new variety doesn't lodge or succumb to a disease. Increasing the yield potential is not as obvious; it is only apparent when compared with other varieties in controlled experiments, but it is arguably more important in the long run.

Some scientists argue that yield potential has not changed over time and all yield improvement was a result of increased stress tolerance (defect elimination). The argument was based on research with maize, but the concept could be generalized to all grain crops. It seems unlikely, however, that the varieties and hybrids in use at the beginning of the high-input era of agriculture had the capacity to produce today's yields without increases in yield potential. I do not believe that defect elimination (considering a lack of stress tolerance to be a defect) by itself could convert maize varieties that yielded roughly 1900 kg ha^{-1} (30 bu acre^{-1}) in 1920 into hybrids that yield more than 19,000 kg ha^{-1} (300 bu acre^{-1}) today. Surely, yield potential must have increased.

There is a general belief that modern varieties are more stress tolerant than older varieties. There is little direct evidence of the change and it is unlikely that breeders purposely selected for stress tolerance. Selecting for yield in a typical field environment that often contained some stress, however, would probably select indirectly for stress tolerance. The varieties that had the highest yield would probably have some tolerance to the stresses encountered in the selection environments. Recent development of drought-tolerant maize hybrids (Cooper *et al.*, 2014; Nemali *et al.*, 2014) focused directly on stress tolerance. Varietal improvement in stress tolerance should result in decreases in yield gaps (potential yield – producer yield) over time. A recent evaluation of changes in yield gaps from 1972 to 2011 for county yields of rain-fed maize in two Midwestern states (Iowa and Kentucky) provided only meagre evidence of decreases in the magnitude of yield gaps (i.e. increases in stress tolerance). There were declines in ten of 47 Iowa counties, but only three of 32 counties in Kentucky (Egli and Hatfield, 2014b). The results were similar for soybean (Egli and Hatfield, 2014a). Finding declining yield gaps in so few counties does not provide strong support for an increase in stress tolerance over time, at least increased tolerance to water stress. It is possible that greater stress in the environment from climate change could have cancelled the improved stress tolerance of the varieties and hybrids, leading to little change in the yield gaps over time. The question of increased stress tolerance in modern varieties or hybrids remains an unanswered question.

Breeding for higher yield potential is usually accomplished by selecting for yield, without considering the plant processes involved in the production of yield. Crop physiologists have been busy for years identifying physiological processes associated with yield, hoping that plant breeders could select for these processes and speed up yield improvement. Except for a few isolated examples, this approach was not successful in the public sector (Egli, 2017, p. 166); we don't know if, or how much, it was used by commercial breeders. Reasons for its failure include a lack of genetic variation for many target traits, identification of traits that were only marginally related to yield and an inability to adequately characterize complex physiological traits on the large number of plants involved in a breeding programme. A more fundamental problem with this approach stems from the complex nature of the yield production process (see Chapters 2 and 3, this volume), making it difficult to increase yield by selecting for a single plant process. Selecting for yield may still be the best approach.

The coming of the biotech era greatly increased our ability to genetically manipulate individual plant processes, raising hopes that a new age in plant improvement was dawning. However, increasing yield by manipulating individual genes turned out to be more difficult than originally expected. There have been many reports of individual genes that were related to yield but translating these findings to producers' fields has not been easy (see Chapter 5, this volume). Biotechnology was used to develop drought-tolerant maize hybrids (Nemali *et al.*, 2014), but it seems that much of this new age of crop improvement still lies in the future. Biotechnology and molecular biology are being used productively as an adjunct to conventional breeding and it seems likely that these constantly improving techniques will eventually increase the rate of yield improvement.

The increase in grain crop yields that started with the advent of high-input agriculture (see Chapter 1, this volume) is usually attributed to a combination of better management and improved varieties. The relative contribution of the two is a subject of much debate and any possible contribution of changes in climatic conditions (positive or negative) is usually ignored. Many agronomists assign roughly half of the increase to improved management and the other half to improved varieties for maize and soybean, but, in fact, the two agents of improvement do not operate independently. Improved varieties may require changes in management practices to express their higher yield potential, making it difficult to isolate breeding and management as independent agents of change.

The classic example of this interdependency was the development of the short-statured wheat and rice varieties that triggered the Green Revolution. The high yield of these varieties was realized only under intensive management including higher rates of N fertilizer (Hessor, 2006, p. 58). A second example is the upright leaf trait that is common in modern maize hybrids. This trait increases yield at higher LAIs, but at low LAIs it reduces yield by decreasing solar radiation interception (see Chapter 3, this volume). Consequently, hybrids with upright leaves produce higher yields only when grown at higher populations (Duncan, 1971).

Although it is difficult to identify the specific contributions from better management and improved varieties, I think variety improvement made a larger contribution to the historical yield increase, especially in the last few decades, than management. As discussed previously, management increases crop yield by improving the crop's environment; as the environment approaches the 'perfect' environment, it becomes more difficult for new management practices to improve yield. It seems that precision agriculture technology has the potential to contribute more to increasing efficiency than to increasing yield, supporting my contention that it is difficult to increase yield by management as the environment approaches the 'perfect' environment.

The diminishing returns spiral associated with management makes it unlikely, in my opinion, that improving management would be able to support a continuous increase in yield for over 100 years. The same logic applies to defect elimination breeding. Each defect that is eliminated makes it more difficult to identify another defect and increase yield. Increasing yield potential, however, has no such limitation, at least not yet. Each new variety released by plant breeders generally has higher yields than its predecessors do and this more or less continuous improvement is the type of improvement needed to support the steady historical yield increase. Improved management contributed to the historical yield increase, but the primary driver was, and will be, in my opinion, improvement in yield potential.

We are making a mistake, however, if we completely discount contributions from improved management. Management, including precision agriculture technologies, contributes to improvements in production efficiency. These improvements may not necessarily increase yield, but they will decrease inputs per unit of yield, improving the producer's bottom line. The new management technologies coming online will no doubt continue this improvement in production efficiency. Management may also play a significant role in minimizing environmental damage and adjusting our production systems to maintain productivity in a changing climate. The potential interactions of big data, artificial intelligence, algorithms and precision agriculture techniques may make larger contributions from management possible in the future.

Selecting the 'perfect' variety or hybrid determines the success of any cropping system. The characteristics of this 'perfect' variety are obvious: appropriate maturity for the location, high yield, the best agronomic traits (e.g. lodging and shattering resistance), necessary disease and insect resistance, and herbicide tolerance traits pertinent to a particular cropping system. University-run variety tests, when available, are excellent sources of this information. It might seem that this selection process is relatively simple and straightforward, but, unfortunately, that is not necessarily true.

Selecting varieties is a never-ending task – new varieties of most grain crops are constantly appearing in the marketplace. These new offerings are superior to the ones they replace. This constant appearance of new, improved varieties results in rapid obsolescence of older varieties. The 'shelf' life of modern maize hybrids or soybean varieties is relatively short; less than 30% of the maize and soybean varieties tested by the University of Kentucky since 2000 remained in the test for 3 years (Fig. 4.1). This rapid turnover of varieties does not necessarily occur in other cropping systems. For example, the tall fescue forage variety 'KY 31', released to producers in 1942, is still the predominant tall fescue variety in production in Kentucky in 2019 (G. Lacefield, University of Kentucky, 2019, personal communication). A grain farmer using a variety that was released in 1942 would not be a grain farmer for very long. The superiority of today's varieties is a major contributor to high yields and the producer must constantly change varieties to capture this superiority.

Variety maturity – length of the growth cycle (planting to maturity) – is an important part of variety selection and there is substantial variation to choose from (see discussion in Chapter 3, this volume). This selection criterion is absolute: the variety must fit into the available growing season whether the growing season is defined by temperature or moisture availability. The variation in the available growth durations provides the flexibility to find a variety that 'fits' into a range of growing seasons, including those that are relatively short. The movement of US maize production north into North and South Dakota and the prairie providences of Canada, for example, was partially associated with

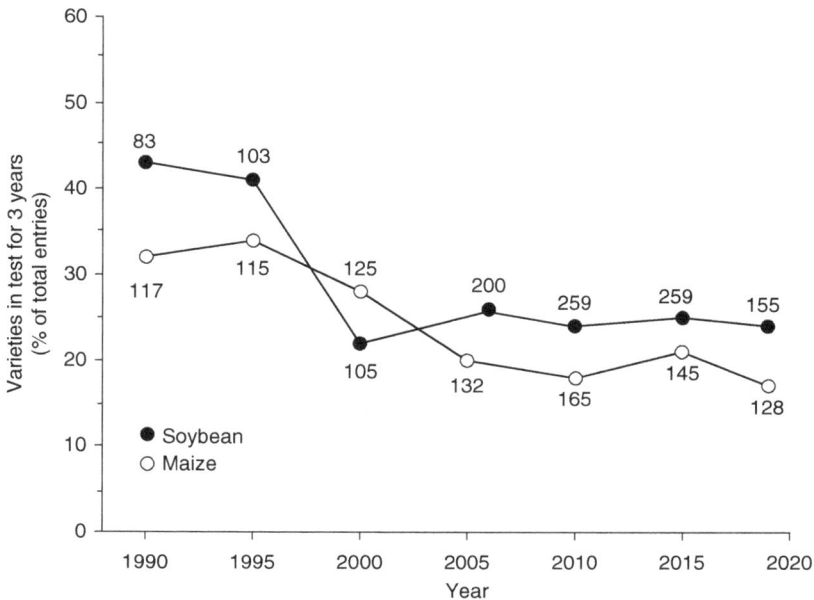

Fig. 4.1. The longevity of varieties and hybrids in the soybean and maize performance tests in Kentucky, USA, 1990 to 2020. Numbers beside the lines indicate the number of varieties. (Data from the Performance Test Bulletins.)

the development of high-yielding hybrids with short total growth durations. The availability of varieties with variable growth durations in most grain crops made it possible to select a variety that used most of the available growing season at most locations. The historical recommendation to use full-season varieties was based on the assumption that yield increases in step with growth duration, but, unfortunately, this assumption is not necessarily true.

Increasing the total growth duration increases the length of the vegetative growth phase, but the length of the reproductive growth phase stops increasing when the total growth duration reaches about 100 to 120 days (see the detailed discussion in Chapter 3, this volume). Yield will not increase if length of the reproductive growth phase is constant unless the rate of growth increases. This is unlikely, however, because a longer total growth duration (later-maturing varieties) often shifts reproductive growth of summer annuals into a less productive environment later in the growing season (temperature and solar radiation are lower). In light of these relationships, the recommendation for the use of full-season varieties is hard to understand unless it is a holdover from the pre-herbicide era when wide rows were needed for mechanical weed control and large vegetative plants (large LAI) were needed to maximize solar radiation interception and yield. Large plants are not needed in modern narrow-row production systems to ensure complete ground cover. Yield should increase in step with the total growth duration until it reaches 100–120 days; beyond that, the primary effect is simply to produce larger vegetative plants, but not necessarily higher yield.

Short-duration varieties provide some unique opportunities to increase productivity or production efficiency (see discussion in Chapter 3, this volume). Short-duration varieties enhance double-cropping opportunities, provide the flexibility to match reproductive growth with the most productive part of the growing season and require fewer inputs in some situations (e.g. less irrigation water or perhaps fewer applications of crop protection chemicals) simply because of their shorter duration. A crop that matures in 100 days will likely use less total water, and hence require fewer irrigations, than a crop that matures in 140 days. The saturation relationship between total growth duration and reproductive growth duration (Fig. 3.9, Chapter 3, this volume) suggests that these efficiencies can be accomplished without sacrificing yield. Short-duration varieties, however, may be more susceptible to stress during vegetative growth. The smaller vegetative plant with a lower LAI makes it more likely that stress could reduce the LAI below the critical level, reducing solar radiation interception, CGR and yield.

Another consideration when selecting a variety is specific versus general adaptation. Is there a variety that is uniquely more productive on your farm or in specific fields on your farm? Perhaps there is a variety that does best on a specific soil type. These are examples of specific adaptation. The opposite of specific adaptation is a variety that produces high yields over a large geographic area, i.e. one that has general adaptation. In short, do varieties have specific or general adaptation? Historically, the goal of most plant breeders was to develop varieties that were adapted to large geographical areas that included a wide range of environmental conditions, i.e. they had general adaptation. Varieties with disease or insect resistance or drought tolerance would perform well in the presence of these stresses, so they could be taken as an example of specific adaptation unless the stress was widespread. Specific adaptation, however, is usually considered a narrower adaptation to environmental conditions (including both above- and below-ground conditions).

Specific adaptation is frequently promoted by seed companies and crop consultants and the concept is often accepted by producers. Is there a variety/hybrid that does well in one field but not in another (ignoring the obvious effects of the presence of specific diseases or insects)? In spite of the relatively widespread belief in specific adaptation, it is difficult to find solid evidence supporting it. Evaluation of several years of soybean variety test data in Kentucky led to the conclusion that the best predictor of soybean variety performance was the 2-year average yield at all locations in the state (Pfeiffer, 1996), not the performance at an individual location, i.e. there was no specific adaptation. Pfeiffer (1996) also found that the use of several varieties reduced the year-to-year variation in yield. Attempts to develop soybean varieties specifically adapted to the wheat–soybean double-cropping production system (i.e. planting late after winter wheat harvest) were not successful, another example of the failure of the specific adaptation approach. It is possible that the application of improved analytics to large data sets characterizing variety performance over a wide range of environmental conditions will reliably identify the variety that will be most productive in a specific environment (specific adaptation), but, in my opinion, such predictive ability is some distance in the future. Extensive implementation of the specific adaptation approach could challenge the seed-supplying capacity of the seed industry. It seems to me that, currently, general adaptation rules. The best variety is the one that shows the highest average yield over locations and years.

In summary, new varieties of most grain crops are constantly made available by the commercial seed industry; utilizing these varieties is the only way to translate genetic improvement (yield, disease and insect resistance, herbicide tolerance) into higher yield in the producer's field. Increasing inputs will not increase yield of older varieties, so constant change of varieties is necessary. Beyond capturing genetic improvement, judicious variety selection, particularly variety maturity, offers opportunities to increase yield by minimizing stress, to reduce inputs without affecting yield (i.e. improving efficiency) and to increase total productivity via double cropping. A successful grain producer will spend some time thinking about what variety(s) to grow.

Plant Population

The goal when selecting a planting rate is to establish a plant population (plants per unit area) that will produce maximum yield. Plant population, at a minimum, must be high enough to ensure maximum solar radiation interception by the beginning of reproductive growth; anything less will reduce yield. If the population is too high, plants may lodge, also reducing yield. Finally, the grower would like to minimize seed costs, especially in this age of high-priced GMO seed. Meeting these seemingly simple, although somewhat conflicting objectives involves consideration of the quality of the planting seed (germination and vigour), the potential characteristics of the seedbed (primarily temperature and moisture levels) and the reproductive characteristics of the species involved.

The first issue that complicates selecting the optimum population is the fact that the producer can only select the seeding rate. The population is determined by the emergence percentage, which (as discussed previously in the 'Planting-Seed Quality' section of this chapter) is a function of the quality of the planting seed, seedbed conditions (temperature, moisture, surface conditions) and planter performance. Planting date, tillage system and the presence of crop residues on the soil surface (no-till and/or cover crops) affect seedbed conditions and the difference between planting rate and emergence percentage. Early planting and/or the use of cover crops and minimum- or no-till systems may result in a cool, wet seedbed that reduces emergence and final stand (see Table 4.1). Higher planting rates along with fungicide and insecticide seed treatments may be necessary to maintain adequate emergence under these less than desirable seedbed conditions.

Expensive biotech seed encourages producers to lower seeding rates to reduce seed costs, but the possibility of lower emergence levels must be considered when making this decision. Lower than expected emergence may require replanting or reduce yield if the seeding rate is set at the minimum level that will produce maximum yield.

The population that will produce maximum yield depends, in part, on the reproductive flexibility (plasticity) of the species; or, to put it another way, how yield per plant responds to changes in plant population. Reproductive flexibility or plasticity was probably an important attribute of undomesticated plants growing in the wild because it allowed the plant to adjust its reproductive output to a wide range of plant populations and environmental conditions. Evolution would select plasticity to maximize reproductive output and most grain crop species have retained this characteristic in modern varieties. Maize and sunflower are two important grain crops that have not retained their plasticity.

Teosinte, the wild ancestor of maize, was very plastic, producing multiple ears on the main stem and ear-bearing tillers. Plasticity, however, has been bred out of modern hybrids; they no longer produce grain-bearing tillers and many produce only a single ear although ear primordia are produced at all nodes below the ear-bearing node (Ritchie *et al.*, 1993). Producers growing open-pollinated varieties before the high-input era probably favoured single-ear types when selecting ears to save for seed for the next crop. Competition for the 'best' ear among farmers in the pre-hybrid era favoured large single ears

(Collins *et al.*, 1965) and single-ear types may have been easier to harvest by hand. Since then plant breeders maintained the emphasis on single-ear types in spite of data showing advantages (e.g. yield stability in drought-prone environments) (Ross *et al.*, 2020) for multiple-ear hybrids.

Selecting a population is relatively easy in species where seed number per plant is flexible (plastic species). Seeds per plant on soybean, canola, the cereals (wheat, barley, rice, etc.) and grain legumes (field pea, common bean, lentil, lupin) decrease as the plant population increases over a wide range, maintaining a constant number of seeds per unit area and yield as shown for wheat and barley (Fig. 4.2) and soybean (Fig. 4.3). If seeds per unit area and solar radiation interception are constant there will be no change in yield over a wide range in population. These species are not sensitive to changes in population because seeds per plant is flexible (plastic). Seeds per unit area and yield will decrease, of course, if the population is too low and solar radiation interception during reproductive growth is reduced (low populations in Fig. 4.2 and 4.3). Once the population is high enough to maximize solar radiation interception by the beginning of reproductive growth, there will be no further changes in yield as population increases, unless a very high population causes lodging (generally associated with reductions in stem diameter) and reductions in yield. Selecting a population for flexible species is relatively easy because yield remains constant over a substantial range in population.

Do not forget that plant population is not the only factor influencing the capacity of the crop community to reach complete ground cover (see Chapter 3, this volume). Other aspects of management (row spacing, variety maturity and planting date) are also important. Higher plant populations are often

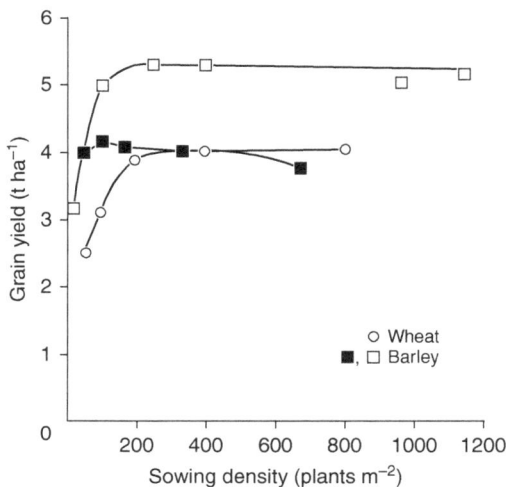

Fig. 4.2. The effect of plant population on yield of wheat and barley. (From Evans, 1993, p. 181.)

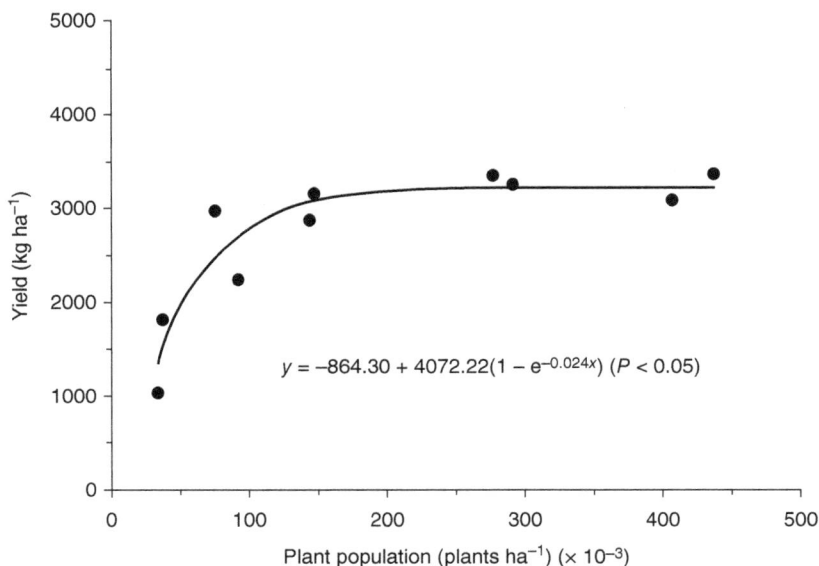

Fig. 4.3. The response of soybean yield to plant population. The variety 'Cavendale Farms CF461' (relative maturity 4.3) was planted on 21 May 2003. (Adapted from Lee *et al.*, 2008. Used with permission from John Wiley and Sons.)

recommended for some grain crop species to compensate for wide rows, smaller plants resulting from late plantings, low-productivity environments or the use of early-maturing varieties.

Managing population is complicated for non-plastic species like maize or sunflower. The ability of these species to adjust seeds per plant is limited by the size of the reproductive structures (florets per ear(s) (maize) or per capitulum (sunflower)). Plants of non-plastic species no longer have the ability to produce more reproductive structures in response to increased photosynthesis per plant at low populations or in highly productive environments. To overcome this limitation and increase seed number in response to higher levels of crop productivity, population must be increased to provide the necessary reproductive structures to increase seeds per unit area (Egli, 2015a, 2019). If the population is too low, the crop will be sink limited, because there will not be enough florets and seeds available to utilize all of the assimilate from photosynthesis and yield will be limited.

The population that produces maximum maize yield is a function of hybrid characteristics (florets per ear (potential kernels per ear), tendency to produce a second ear and kernel size) and the productivity of the environment (the yield level). A hybrid with a small ear will require a higher population than a hybrid with a large ear at any yield level. For example, a hybrid yielding $1568 \, g \, m^{-2}$ ($250 \, bu \, acre^{-1}$) with a $300 \, mg$ kernel and 800 potential kernels ear^{-1} (16 rows and 50 potential kernels per row) will require at least $6.5 \, plants \, m^{-2}$ ($26,316$

plants acre^{-1}). A hybrid with a smaller ear (576 potential kernels ear^{-1}) (16 rows and 36 potential kernels per row) will require at least 9.1 plants m^{-2} (36,842 plants acre^{-1}) (a 40% increase) to produce the same yield. The yield, kernel size and kernels per unit area stayed the same in this example, so reducing ear size required more plants to produce the required number of kernels. Decreasing kernel size will affect plant population in a similar manner. Decreasing kernel size from 300 to 260 mg kernel^{-1} in the previous example (yield of 1568 g m^{-2} and 800 potential kernels ear^{-1}) requires more kernels per unit area and the plant population needed to produce those kernels (minimum population) must increase from 6.5 plants m^{-2} (26,316 plants acre^{-1}) with a 300 mg kernel to 7.6 plants m^{-2} (30,769 plants acre^{-1}) with a 260 mg kernel (a 17% increase). It takes more small kernels to produce the same yield than it does large kernels, so with the same ear size, a higher population is needed. The yield component equation (Eqn 3.3, Chapter 3, this volume), yield = seed number × seed size, provides the basis for these calculations. Interestingly, investigations of population–yield relationships in maize, and there are many, rarely consider ear size or kernel size of the hybrids tested.

Aborted kernels at the tip of the ear at maturity are an indication that all of the florets were not needed to convert the productive capacity of the maize plant into yield. Aborted kernels at the tip of the ear indicate that the crop was not sink limited, a condition that occurs when there are not enough florets (kernels) to use all the available photosynthate. On the other hand, ears with no aborted kernels at the tip (often prized by growers as an indication of high yield) could also be an indication of a sink limitation and lost yield, i.e. yield would have been higher if there were more florets (bigger ears or higher populations). Stress (reduction in photosynthesis) during Murata's Stage II, the critical period for kernel number determination, can also cause tip kernels to abort and, in this case, yield is lost. If the population is ultra-high relative to productivity, or if there is excessive variation in the time of seedling emergence or in plant-to-plant spacing, photosynthesis on some plants may be reduced to the level that none of the florets develop into kernels, resulting in barren plants.

The maize plant would gain some plasticity if the plant produced a second ear or if the number of florets per ear increased in response to favourable environmental conditions during ear development or to lower populations (more photosynthesis per plant in both cases). Unfortunately, only a few hybrids produce second ears. Scientists demonstrated that N or water stress during ear development reduced florets per ear, but studies with population generally show little effect. I know of no data showing that favourable conditions during ear development will increase ear size (number of florets per ear). It seems that even the so-called flexible-ear types do not significantly increase the plasticity of the maize plant.

The presence or absence of plasticity in a grain crop species determines the role that plant population plays in determining productivity. As plant breeders increased maize productivity during the hybrid maize era (Fig. 1.1, Chapter 1, this volume), ear size did not increase in step with productivity, so it was necessary to continually increase plant population to have enough florets (potential kernels) to translate productivity into yield and avoid a sink limitation (Egli, 2015a). This simple explanation explains why maize population increased from roughly 10,000 to 20,000 plants ha^{-1} (4000 to 8000 plants acre^{-1}) for open-pollinated varieties before the advent of hybrids to more than 86,000 plants ha^{-1} (35,000 plants acre^{-1}) today in highly productive environments. Yield contest winners report populations of 128,000 plants ha^{-1} (50,000 plants acre^{-1}) or higher to support their ultra-high yields. Future yield increases will require even higher populations, but there is a limit to how many plants can occupy a given length of row. Row spacing will probably have to decrease or a twin-row planting pattern adopted to accommodate yields of the future unless weight per kernel, ear size (florets per ear) or ears per plant increases.

In contrast, soybean, a plastic species, has experienced declining populations over time. Planting rates recommended for soybean in the middle of the last century were traditionally much higher than needed for maximum yield; often only 50% emergence was needed to produce adequate populations. Planting seed at that time was relatively cheap and it was rarely treated with fungicides, so high planting rates provided insurance against reductions in seedling emergence. If stress in the seedbed reduced emergence, the final population would still be high enough to maximize yield. Higher seed costs in the current biotech era stimulated interest in lower seeding rates to reduce production costs. This decrease in population did not affect yield and it occurred when soybean yield was increasing substantially, providing a stark contrast with maize where higher populations are needed to drive yield increases. Soybean seeding rates must allow an adequate cushion to compensate for reduced emergence, unless you are a high-stakes gambler, because emergence is almost always less than 100%. This comparison of maize and soybean provides a vivid illustration of the importance of plasticity in determining the response of grain crop species to changes in population.

Seeds per unit area and yield of non-plastic species may be reduced if there is variation in the spacing between plants in the row (spatial variation) or in the time of emergence of individual seedlings (temporal variation). Dominant plants, those with wider in-row spacings (more area per plant) or from early-emerging seedlings, have access to more solar radiation and grow faster than dominated plants, those with narrower in-row spacings (less area per plant) or late-emerging seedlings, that grow slowly because they are exposed to less solar radiation per plant. Number of seeds on the dominant plants of plastic species increases and compensates for the reduction of seeds on the dominated plants; consequently, seeds per unit area and yield are not affected (Egli, 1993b) (Table 4.2). This relationship holds as long as the variation in the distribution of plants does not reduce the interception of solar radiation (unlikely unless spatial variation creates large skips in the row).

Seeds per unit area and yield of non-plastic species will be reduced when the dominant plants cannot produce enough seeds to compensate for the

Table 4.2. Effect of delayed emergence on seeds per plant and yield of field-grown soybean. (Adapted from Egli, 1993b.)

Treatment[a]	Seeds plant^{-1}	Yield[b]	
		g plant^{-1}	g m^{-2}
Control	75[b]	12.8	297[c]
4-day delay			
Control	86	15.2	
Delay	52	9.0	
Average	68	12.1	304
7-day delay			
Control	96	16.4	
Delay	39	6.5	
Average	69	11.4	294

[a]Control seeds were planted at the same time in all treatments. Every other seed in the row was planted 4 or 7 days after the control in the delay treatments.
[b]Average of 2 years.
[c]Combined yield of all the plants in each plot. There was no statistically significant difference in total yield among treatments ($P = 0.05$).

loss by the dominated plants. This failure occurs when the number of florets on the ear is not large enough to accommodate all of the seeds needed on the dominant plant to offset the reductions on the dominated plants. Yield reductions will depend on how much variation there is in spacing or time of emergence and how much excess capacity there is on the ear. If all of the florets on the ear are needed to match the yield level of the crop in a perfectly uniform stand (coefficient of variation (CV) = 0), the ear has no excess capacity (no unfilled seeds at the tip), so any spatial or temporal stand variation will reduce seeds per unit area and, presumably, yield (the 8.2 plants m^{-2} (33,198 plants acre^{-1}) treatment in Fig. 4.4a). If, however, only a portion of the florets is needed to match the yield level in the uniform stand (population of 9.1 plants m^{-2} (36,842 plants acre^{-1}) or higher in Fig. 4.4a), the ear has excess capacity which can be utilized to increase the number of seeds on the dominant plants, preventing some yield loss as variation increases. When the excess capacity is exhausted, seeds per unit area and yield will start to decline. The same concept applies to variation in ear size at a constant population (Fig. 4.4b). If ear size is just large enough to match the yield level of the uniformly spaced (CV = 0) crop, seeds per unit area starts decreasing as soon as spatial variation exists (CV > 0) (576 kernels ear^{-1} in Fig. 4.4b). Larger ears provide excess capacity and delay the decrease in seeds per unit area as variation increases. The response to increasing spatial variation or temporal variation is the same when the excess capacity is created by population (Fig. 4.4a) or ear size (Fig. 4.4b). In both scenarios, yield reduction depends upon how much variation there is in spacing or time of emergence (magnitude of the CV) and how much excess capacity there is on the ear.

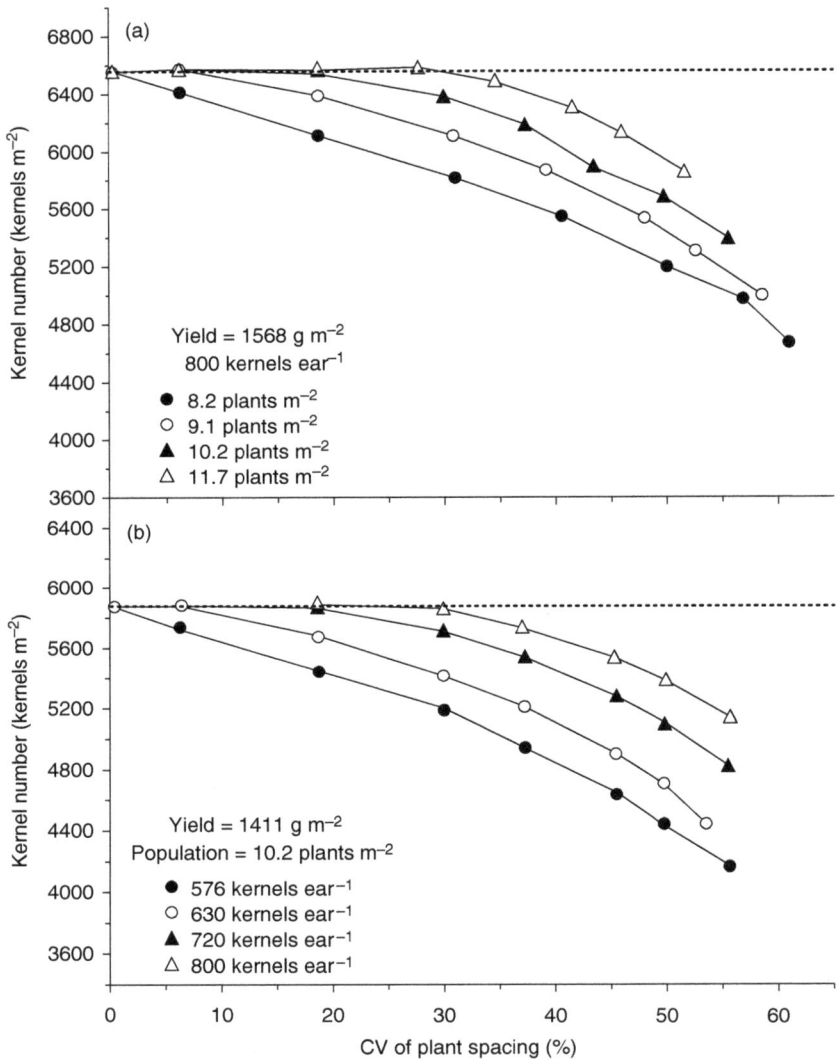

Fig. 4.4. The effect of variation of the spacing between maize plants in the row on kernel number (kernels m^{-2}). The data were produced by KNMAIZE, a simulation model of kernel set in maize. The coefficient of variation (CV) is a measure of the variation of the in-row spacing. Population varies in (a) with a constant ear size, while ear size varies with a constant population in (b). (D.B. Egli, 2021, unpublished results.)

Uniformity of spacing and time of emergence are important for non-plastic species, but populations higher than that needed for maximum yield will provide excess ear capacity and reduce the effect of spatial and temporal variation on seeds per unit area and yield.

Perfect spatial uniformity requires a planter that places each individual seed at precisely the same distance from the preceding seed, as well as 100% germination and emergence. The 'missing' plants will reduce uniformity if seed germination and emergence are less than 100%. Variation in the time of emergence is related to variation in planting depth and the moisture content and temperature of the soil surrounding each individual seed. Recent advances in planter design reduced the variation in spacing between individual plants and probably contributed to higher and less variable emergence.

As discussed previously in this chapter, the quality of the planting seed (standard germination and vigour) influences performance of the seed and seedling in the field. High-vigour seed generally emerges faster and may have higher final emergence, especially when there is stress in the seedbed. High-vigour seed also produces less variable emergence. Interestingly, the uniformity of emergence of maize and soybean seedlings decreased as the time to emergence increased (Egli *et al.*, 2010; Egli and Rucker, 2012). Consequently, low soil temperatures, often associated with early plantings, that delay emergence may decrease uniformity even if high-vigour seed is planted. This decrease in uniformity may reduce the yield advantage of early-planted maize. The decrease in temporal uniformity would not affect the yield of plastic species such as soybean.

Reproductive plasticity governs the response to spatial and temporal non-uniformity and the response to variation in population. How much can an individual plant increase the number of seeds it produces when exposed to a more favourable environment? If the potential increase is large (plastic or flexible species), yield is constant over a wide range of population and spatial and temporal non-uniformity. If, however, the potential increase is small (non-plastic species), yield is sensitive to population and to spatial and temporal variation. The perfectly uniform stand (spacing and time of emergence) does not exist in production fields, but the stand does not have to be perfectly uniform to produce maximum yield. Even non-plastic species can tolerate some non-uniformity if the population is high enough to allow compensation between dominant and dominated plants (Fig. 4.4).

In summary, managing population is relatively easy for plastic species that produce the same yield over a range of populations. Uniformity of spacing and time of emergence are not important for these species; however, there must be enough plants to ensure complete ground cover and maximum interception of solar radiation by the beginning of reproductive growth. Managing population of non-plastic species is more complicated. The minimum population producing maximum yield in these species depends on their reproductive characteristics (ear size (florets per ear), ears per plant, seed size) and the productivity of the environment (yield level). Interestingly, research on the response of maize yield to population seldom considers ear size or seed size, even though recent theoretical modelling work supports their importance (Fig. 4.4) (Egli, 2019). Non-plastic species are sensitive to a lack of spatial and temporal uniformity, so precision planting is an important component of their production system.

Planting Date

There is no simple, well-defined rule that tells a producer when to plant his crop. Obviously, the choice of a planting date must meet the minimum requirement of avoiding low-temperature damage to the plants at the beginning (reduced emergence due to low soil temperatures or freeze injury to emerged seedlings) and end (exposure to freezing temperatures before maturity) of the crop's life cycle in temperate climates. Planting date in tropical climates with distinct wet and dry seasons must position growth of a non-irrigated crop in the wet season. Planting date is more flexible in tropical climates with no temperature or moisture limitations. Multi-cropping systems, e.g. double-cropping soybean after winter wheat in temperate climates, may force selection of a non-optimum planting date for one of the crops. Beyond these obvious restrictions, the mechanistic principles underlying the choice of a planting date are not always obvious.

Planting date is a management practice that can be easily manipulated without necessarily incurring extra operational or equipment costs. Its importance and the relative ease of conducting planting date experiments made it a favourite topic of researchers over the past 100 years or so. The first planting date experiment with soybean that I am aware of, for example, was conducted in Tennessee in 1908 (Mooers, 1908) and researchers are still publishing results from planting date experiments 107 years later (Boyer *et al.*, 2015). A survey of the refereed literature in 2009 (Egli and Cornelius, 2009) found 28 multi-year soybean planting date experiments published between 1960 and 2005. These published experiments were probably only a fraction of the experiments conducted during this period. I suspect that similar publication records could be documented for maize. In spite of over 100 years of investigations with soybean and other crops, the fundamental basis for the response of yield to planting date is still not clearly understood. One could argue that it should not take that many years of often intensive experimentation to understand this phenomenon. This failure, in my opinion, is a result of the tendency of researchers to simply document the response, without asking why the response occurred.

Historically, summer annual grain crops (e.g. maize and soybean) were not planted until soil temperature reached a favourable level ($\sim 10^\circ$C (50°F) for maize and 15.6 to 18.3°C (60 to 65°F) for soybean) for germination and seedling emergence. Planting in cooler soils was not recommended unless weather forecasts predicted a steady increase in soil temperature in the days following planting. Soil temperature is not an important consideration in modern production systems, resulting in generally earlier planting dates in the maize belt of the USA (Kucharik, 2006). Successful stand establishment in these earlier planting dates may be partially due to better seed pesticide treatments that protect the slower germinating and emerging seedlings, and to hybrids and varieties with better early-season stress tolerance (Abendroth *et al.*, 2017). Other aspects of cropping systems that could contribute to successful early plantings include earlier spring warm-ups as a result of a changing climate, producers' desire to exploit the higher yield potential observed in early plantings, larger farming units requiring a longer planting season coupled with fear that weather-induced planting delays could eventually lower yields, and other changes in technology such as improved herbicides and improvements in planter design (Abendroth *et al.*, 2017).

Maize and soybean yields generally decline as the planting date is delayed (Figs 4.5 and 4.6), but the response varies from year to year, indicating a substantial effect of weather conditions. The variation among years or locations in Fig. 4.6 is probably due to variation in the amount and distribution of rainfall during critical growth stages. Temperature and solar radiation are usually less variable among years and locations than rainfall and their contribution to the year-to-year yield variation is probably not as important. The average response across experiments and locations reflects the response to average weather conditions and it is this average that, in the absence of accurate long-range (5 to 6 months) weather forecasts, must be used for planning purposes. Most producers can remember the year when they planted late and produced higher yields than earlier plantings as a result of a more favourable rainfall distribution. These exceptions can happen, but the average response (lower yields in late plantings) is much more likely.

The timing of the initiation of the decline in soybean yield seems to have changed in recent years. In experiments conducted between 2006 and 2011, the rapid decline in soybean yield started in early May (Fig. 4.6) (Knott *et al.*, 2019) and maximum yield occurred in mid-April. This advantage for early

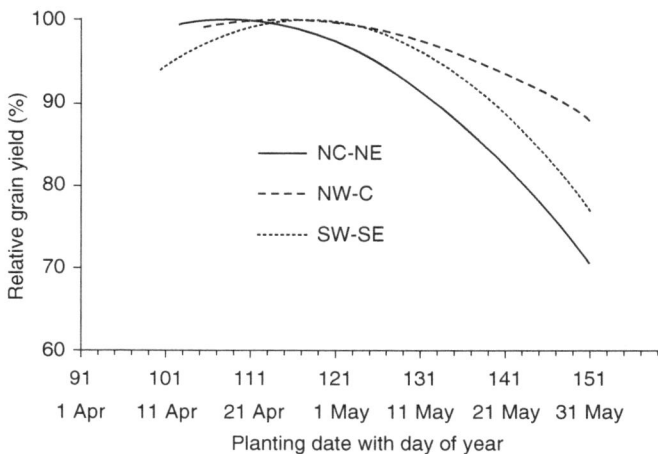

Fig. 4.5. Response of maize yield to planting date in Iowa, USA. The response was averaged within three areas of the state: NC-NE, north central and north-east; NW-C, north-west and central; SW-SE, south-west and south-east. (From Abendroth *et al.*, 2017. Used with permission from John Wiley and Sons.)

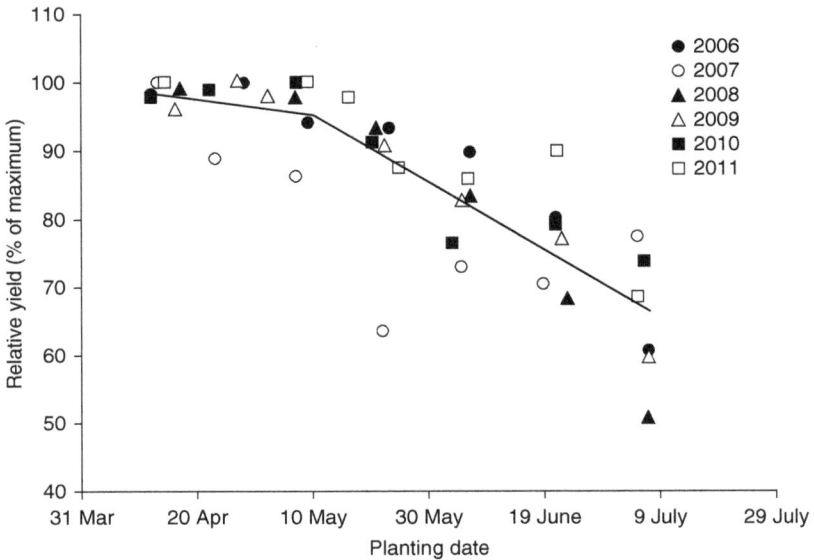

Fig. 4.6. Effect of planting date on yield of a Maturity Group IV soybean variety in Kentucky, USA, 2006 to 2011. (From Knott *et al.*, 2019. Used with permission from John Wiley and Sons.)

planting of soybean occurred in other recent experiments in several states. Yield reached a maximum in mid- to late April in experiments with maize conducted from 2006 to 2009 (Fig. 4.5) (Abendroth *et al.*, 2017), a response similar to the soybean experiments.

Older experiments show a completely different response. Soybean yield did not change as planting was delayed from late April until approximately 1 June, 60 days after 31 March (Fig. 4.7) (Egli and Cornelius, 2009), in a summary of experiments conducted in the US maize belt between 1960 and 2005 (only one of the nine experiments was conducted after 1990). Similar responses in this time period were reported by researchers in other countries (e.g. China, Australia, Argentina and Italy), although ultra-early planting dates that are possible in warmer climates produced lower yields. During this era, maize yield reductions started when planting occurred after 1 May, so it declined at earlier planting dates than soybean (Fig. 4.8) (Scott and Aldrich, 1970). This differential response was used to explain why maize was planted before soybean.

The response of these two crops to planting date has changed over time and it is unlikely that the change is just a response to random variation in weather conditions from year to year, given that multiple experiments covering several years document both responses. The reasons behind the change are not clear, but the earlier warm-up in the spring, technology that allows stand establishment at lower soil temperatures and the development of more stress-tolerant varieties and hybrids all could be involved. Modern data suggest that

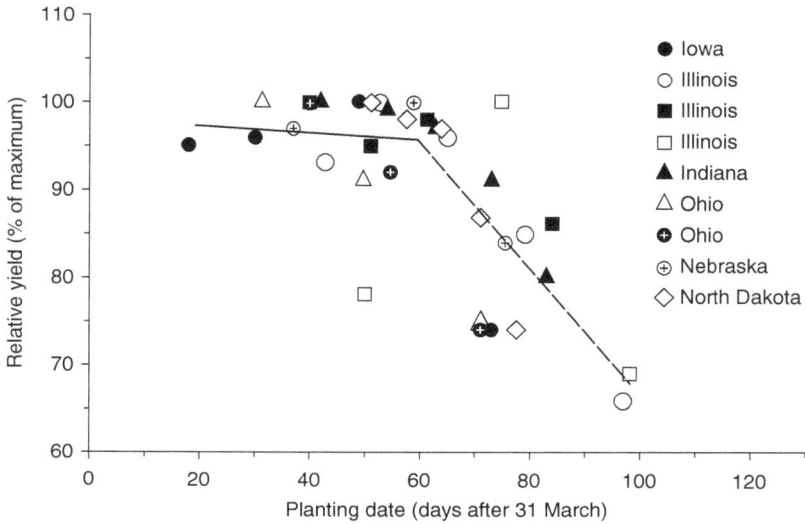

Fig. 4.7. Effect of planting date on soybean yield in the Midwestern USA ($n = 35$). Includes data published in refereed journals between 1960 and 2005 (only one source was published after 1990). (From Egli and Cornelius, 2009. Used with permission from John Wiley and Sons.)

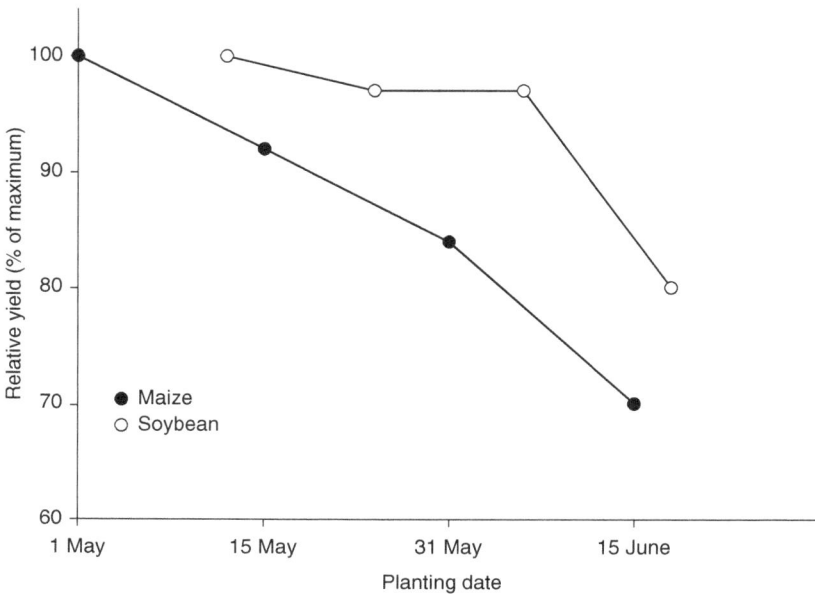

Fig. 4.8. Response of maize and soybean yield to planting date in the 1960s. (Adapted from Scott and Aldrich, 1970, p. 54.)

planting of both crops must occur in April for maximum yield, which probably partially explains the trend for earlier planting in the maize belt and the recommendation by some that soybean should be planted before maize. Maize yield is more sensitive than soybean yield to reductions in population below the target population and to variation of in-row spacing between plants and time of emergence of individual seedlings. Both spatial variation and temporal variation among plants are more likely to occur in early plantings, so planting the relatively insensitive soybean first would limit effects on yield.

Why do yields decrease when planting is delayed? Answering this question could lead to better management decisions, but it requires a mechanistic understanding of the planting date response. I will focus on soybean to investigate this question, but much of our discussion will relate, equally well, to maize and probably other summer-grown grain crops. The response of winter annuals to planting date is complicated by the period of dormancy between cessation of growth in the autumn and its resumption in the spring and will not be discussed.

Planting date, along with variety maturity, positions the growth cycle of the crop in the growing season, determining when critical reproductive growth stages occur and, therefore, the weather conditions they will likely encounter. The planting date that exposes the crop to the most favourable conditions during reproductive growth should have higher yields than planting dates that expose it to less favourable conditions. The best environment would have high levels of solar radiation, temperatures that accommodate maximum photosynthesis and a long seed-filling period, and adequate, well-distributed rainfall. It may be difficult, however, to combine low temperatures that favour a long seed-filling period with the higher temperatures that maximize photosynthesis. Adequate rainfall depends upon its distribution, the soil water-holding capacity and the ET_0 (atmospheric demand for water) as discussed in Chapter 2 (this volume).

Average solar radiation starts to decrease after the summer solstice (approximately 21 June in the northern hemisphere and 21 December in the southern hemisphere) (Fig. 4.9), while the decline in average air temperatures starts roughly 1 month later. Average monthly rainfall declines during summer and into the autumn in many Midwestern climates (see Fig. 2.10 for additional examples). There is substantial year-to-year variation in this pattern, which probably accounts for the variation in the response to planting date across years. The variation in rainfall is probably more important than the variation in solar radiation or temperature.

Delaying planting tends to shift reproductive growth of soybean into an environment with lower radiation, lower temperatures and less rainfall on the average (Fig. 4.8 and Table 4.3). The critical period for seed number determination (Murata's (1969) Stage II, growth stage R1 to R5) of very early (Maturity Group I) varieties shifted from late June/July to August when planting was delayed from mid-May to late June in Kentucky. The delayed planting of adapted

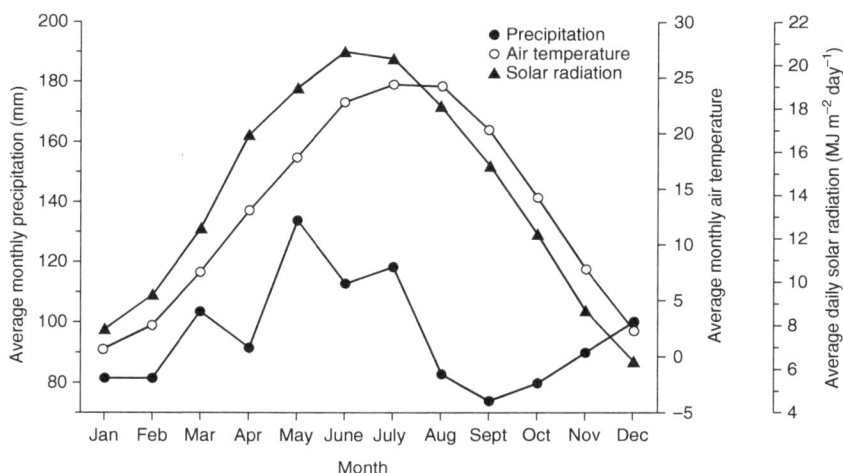

Fig. 4.9. Seasonal trends in long-term average weather conditions at Lexington, Kentucky, USA.

Table 4.3. Planting date and variety maturity position reproductive growth in the growing season. (Adapted from Egli and Bruening, 2000.)

Variety maturity	Date of R1[a]	Date of R5[a]	R1 to R5 (days)	Date of R7[a]	R5 to R7 (days)	Total duration (days)[b]
Maturity Group I						
18 May[c]	26 June	22 July	26	22 Aug	31	96
25 June[c]	28 July	18 Aug	21	18 Sept	31	85
Maturity Group III						
18 May[c]	3 July	8 Aug	36	9 Sept	32	114
25 June[c]	3 Aug	28 Aug	25	2 Oct	35	99

[a]Growth stages, Fehr and Caviness (1977).
[b]Days from planting to growth stage R7.
[c]Planting date.

varieties (Maturity Group III) positioned Stage II even later in the growing season. The late-planted environment is less productive (lower solar radiation, lower temperatures and often less rainfall) than the environment associated with earlier plantings (warmer with higher levels of solar radiation and higher average rainfall), so it is not surprising that yields decline. The response of rain-fed soybean to planting date tends to vary from year to year, which is probably a result of variation in rainfall patterns and amounts. A dry period in late June or early July followed by resumption of rain could easily result in higher yields from the late plantings. Although rainfall influences the response to planting date, irrigation does not usually eliminate the decline in yield from delayed plantings, so the availability of water is not the only cause of lower yields. It is not yet possible to adjust planting date to take advantage of rainfall variation during the growing season (i.e. accurate long-range forecasts are not yet available), so we must schedule plantings based on the average conditions and accept the fact that sometimes it will not produce maximum yield.

Delaying planting shortens the crop's life cycle (Table 4.3) and some think this shortening is responsible for lower yields. This argument is based on the widely held belief that yield increases as the length of the crop's growth cycle increases because it exposes the crop to more accumulated solar radiation. The weakness of this philosophy was discussed at length in the 'Variety Selection' section of this chapter. Seed yield is produced during reproductive growth, so the relative changes in the duration of vegetative and reproductive growth are important when assessing the relationship between shorter growth durations and yield. Delayed planting of soybean shortened the total growth cycle, the vegetative and reproductive growth phases (Table 4.3). Delaying planting of soybean from 18 May to 25 June shortened Stage II (R1 to R5) by 5 days (Maturity Group I) to 10 days (Maturity Group III) (Table 4.3). Interestingly, there was no consistent shortening of the seed-filling period (R5–R7); in fact, it was slightly longer in the delayed plantings. The shorter vegetative growth period in delayed plantings reduces plant height, the number of nodes on the plant, maximum vegetative weight at the end of vegetative growth and maximum LAI. The smaller plants with fewer nodes could lead to reduced pod and seed numbers per plant and may have required higher populations (Egli, 2015b).

Smaller plants with lower LAIs can reduce solar radiation interception and yield, so narrow rows and higher populations may be needed with delayed planting to ensure maximum solar radiation interception by flowering (growth stage R1). In fact, the yield response to narrow rows is much more consistent in late plantings (e.g. planting soybean after winter wheat is harvested) than in early plantings, further demonstrating the importance of solar radiation interception.

So far, our discussion has focused on soybean, but maize yield also declines as planting is delayed (Fig. 4.5) and, in fact, the declines of these two very different species are relatively similar. This similarity is consistent with the conclusion that changes in the environment contribute more to the decline in yield than changes in plant characteristics. Control by the environment could explain the higher yields from very early plantings, which would shift reproductive growth of both maize and soybean into a more favourable environment.

In summary, the reduction of yield with delayed planting seems to be a result of shifting reproductive growth into a less favourable environment rather than changes in plant characteristics. A less favourable environment is consistent with the more or less steady average decline from very early to very late plantings (Figs 4.5 and 4.6) (i.e. solar radiation gets consistently less with each week's delay). Development of varieties specifically adapted to late planting (especially double cropping after winter wheat) has not been very successful, which is also consistent with environmentally induced yield reductions and suggests that little can be done to reduce the effect of delayed planting on yield. Interestingly, manipulating planting date may prove to be useful to reduce the effects of climate change on grain crop yields. Perhaps planting dates can be adjusted to shift crop growth, especially reproductive growth, into a less stressful environment.

Row Spacing

Row spacing – the distance between adjacent rows – is a fundamental characteristic of all grain crop production systems, except when cereals are broadcast seeded. Row spacing varies among crops from the relatively narrow rows associated with cereal crops to the wider rows (0.76 m (30 in)) found in maize, sorghum and soybean production systems. Over time, changes in production technology (primarily the availability of herbicides) made it possible to narrow the rows in most crops. The basic principles that govern the crop response to row spacing are the same for all crops, but the row spacing chosen by the producer is dependent upon crop species, production practices, plant characteristics, and planting and harvesting technology. The long-term trend for the use of narrower rows in maize and soybean suggests that narrow rows are, in fact, superior, but they could not be used until the appropriate technology was available.

Row spacing has been a favourite topic (along with planting date and plant population) of grain crop researchers for many years. One of the first published reports of research on row spacing in soybean appeared in 1939 (Wiggans, 1939) and they were still appearing in 2020 (Schmitz *et al.*, 2020), 80 years later. The results of these experiments have not changed over the 80 years – narrow rows produce the highest yield. It is discouraging, from a scientific viewpoint, that we are still studying the effect of row spacing on yield after all these years; will we ever have enough knowledge to confidently predict the response or do we just need fresh data to satisfy producers?

The row spacing in a producer's fields is dictated by the spacing that produces maximum yield and the production technology available. In fact, historical changes in row spacing were often driven by changes in production technology. Row spacing in maize and soybean in the USA decreased from the more or less standard 1 m (40 in) at the beginning of the intensive agriculture era in the 1930s to 0.76 m (30 in) or less today. Currently more than half of the soybean crop is grown in rows less than 0.76 m (30 in) wide in many soybean-producing states in the USA, while 0.76 m (30 in) is the most common row width for maize (NASS, 2020).

Before the dawn of intensive agriculture, horses were the primary source of motive power, so rows had to be wide enough to accommodate them. Rows also had to be wide enough for mechanical cultivation for weed control in the pre-herbicide era. Mechanization replaced horses and the availability and continual improvement of herbicides eventually eliminated the need for mechanical cultivation and wide rows, so row spacing could be adjusted to maximize productivity within the restraints imposed by planting and harvesting equipment. The availability of equipment for harvesting narrow rows may still limit maize row spacing.

Adjusting row spacing is primarily a matter of maximizing interception of solar radiation by the crop within the limits imposed by the technology associated with each individual cropping system. Row spacing must be narrow enough to maximize solar radiation interception during reproductive growth or yield will be reduced (Fig. 3.2, Chapter 3, this volume). Plant population, plant size and plant characteristics (e.g. vertical versus horizontal leaves, or the ability to tiller or branch (degree of plasticity)) will influence the row spacing needed to reach this goal. For example, late-planted soybean in a double-cropping system after winter wheat requires narrower rows because the plants are smaller with less leaf area (Table 4.4). For this reason, the yield response of soybean to narrow rows is usually larger and more consistent in double-cropped systems than in systems with earlier planting dates. The same relationship applies to the use of early-maturing varieties that produce smaller plants because they have less time for vegetative growth (Table 4.3). Varieties or hybrids with vertical leaves may require narrower rows than those with horizontal leaves for maximum solar radiation interception. The question 'will narrowing rows increase yield?' can be replaced with the question 'will narrow rows increase solar radiation interception during reproductive growth?'. If not, it is unlikely that narrowing the rows will increase yield.

Rows can be narrowed enough in some crops (e.g. soybean) to maximize solar radiation interception very early in vegetative growth. Does this early increase in interception have any effect on yield? Usually there is no direct effect on yield – yield is determined primarily by the solar radiation intercepted during reproductive growth (see Chapter 3, this volume). Early ground cover could, however, lower the costs of production by inhibiting weed growth, improving weed control and possibly decreasing the number of herbicide applications.

Table 4.4. Response of soybean yield to row spacing in a late-planted double-cropping system. (Adapted from Herbek and Bitzer, 1988.)

Row spacing (m)[a]	Yield (kg ha^{-1})[b]	Percentage increase[c]
0.20–0.25[d]	2251[e]	40
0.41–0.51	2083	29
0.76–0.81	1672	–

[a]All plantings occurred after mid-June, simulating double cropping after a winter wheat crop.
[b]Average of 4 years (1972–1974, 1976).
[c]Percentage increase over 0.76–0.81 m rows.
[d]Row width in in = row width in m × 39.37.
[e]Yield in bu acre^{-1} = yield in kg ha^{-1} × 0.01488.

From the viewpoint of weed control, early ground cover is a plus and could indirectly increase yield if weed control improved.

Early ground cover may also increase total water use by the crop (total ET). The soil surface is usually dry and soil evaporation is low, so covering the soil with leaves increases transpiration because the leaf area is larger and the roots have access to water below the surface. Thus, ET usually increases in parallel with LAI until leaves completely cover the soil (see Chapter 3, this volume). Narrowing the rows and increasing early ground cover often increases ET during early vegetative growth, but once the crop reaches complete ground cover, row spacing has no effect on water use. The effect of increased ET from early ground cover on yield will depend greatly on water availability during the growing season. If the amount and distribution of rainfall are adequate, there will probably be no effect. If water is limiting, the excess water used during early vegetative growth may cause the crop to run out of water during repro- ductive growth, thereby reducing yield. Growing crops in wide rows to reduce early water use is a strategy sometimes used in climates where water deficits occur regularly during reproductive growth. The 'saved' water reduces stress during reproductive growth and minimizes the yield reduction.

Narrow rows are often touted for increasing water-use efficiency by decreasing soil evaporation which, the argument goes, should increase the water available for transpiration. This is not true for several reasons. First, soil evaporation is often low because the soil surface is usually dry, so decreasing what is already a small proportion of ET may not be very important. Second, it is not clear why increasing the proportion of the total ET that is transpir- ation is more efficient. Using the concept of efficiency (output per unit input) to evaluate the benefits of early ground cover does not help us understand its effect on yield. The total water use by the crop (ET) along with yield determines the water-use efficiency (yield/total ET), not the distribution between soil evap- oration and transpiration. 'Efficiency' is often used to describe the functioning of crop growth processes and the production of yield, but, as in this case, it rarely improves our understanding of the system.

The value of early ground cover associated with narrow rows can be posi- tive (better or cheaper weed control) or negative (excess water use leading to stress). The advantages of better weed control probably, year-in and year-out, outweigh the disadvantages of excess water use in humid climates with rea- sonably good precipitation distributions. The opposite is probably true in cli- mates where the crop typically runs out of water during the later stages of reproductive growth. In the absence of these considerations, i.e. with perfect weed control and no water stress, complete ground cover before the beginning of reproductive growth has no effect on yield.

The combination of row spacing and plant population determines the space between plants within the row. As population increases with no change in row spacing, the spacing between plants gets progressively smaller and eventually limits population. This limitation is not an issue in flexible or plastic species (e.g. soybean) where yield increases do not require an increase in population, but it will eventually be a problem in non-flexible species (e.g. maize) that can only increase seed number and yield by increasing population (see discussion in the 'Plant Population' section of this chapter). For example, a 31,355 kg ha^{-1} (500 bu acre^{-1}) maize crop will require a population of 22.8 plants m^{-2} (92,368 plants acre^{-1}) (assuming a 275 mg seed and a maximum of 500 kernels per ear and one ear per plant), which results in 5.8 cm (2.3 in) between plants (centre to centre) in 0.76 m (30 in) rows – a continuous wall of maize plants unless the diameter of the stem is less than 5.8 cm (2.3 in). Granted, this is an extreme example (although contest-winning yields have exceeded this level), but it illustrates the problem. Higher and higher maize yields will eventually require rows narrower than the traditional 0.76 m (30 in) to accommodate the higher populations, although increasing ear size, ears per plant or kernel size could offer a temporary respite.

The use of a twin-row planting arrangement could also alleviate the high population problem in maize. In the twin-row system, two rows (twin rows) are arranged in a relatively narrow spacing (e.g. 0.15 to 0.20 m (6 to 8 in)) and the distance to the next set of twins is larger, perhaps equal to the normal 0.76 m (30 in) spacing. In the high yield example in the previous paragraph, the spacing between plants in the row would increase to 11.4 cm (4.5 in) (assuming the twins were 0.76 m (30 in) apart). One advantage of the twin-row approach is that planters that plant maize and soybean in a twin-row configuration are commercially available and maize in twin rows can be harvested with standard 0.76 m (30 in) row harvesting equipment. Twin rows are often touted as a way to produce higher yields, but most research does not support this claim. Twin rows, like narrow rows, will provide higher solar radiation interception early in vegetative growth and the effects would be the same as we discussed previously for narrow rows.

Row direction is not related to row spacing, but occasionally questions about its relationship to yield surface. Does row direction (east–west or north–south or somewhere in between) affect the yield of grain crops? The answer is an emphatic NO! It is true that row direction can influence the amount of solar radiation interception before the crop reaches complete ground cover. East–west rows are roughly perpendicular to the direct beam radiation from the sun at midday in temperate regions so they will probably intercept more solar radiation than north–south rows that are more or less parallel to the sun's rays at midday. Consequently, plants in east–west rows will grow faster before ground cover is complete. Once the crop achieves complete ground cover, however, row direction has no effect on solar radiation interception, so yield will not be affected. This argument assumes complete ground cover by the beginning of reproductive growth; if this criterion is not met, row direction could have an effect, but yield would be less than if complete ground cover occurred before the beginning of reproductive growth. In a sense, this lack of an effect of row direction is unfortunate because, in many situations, row direction could be easily manipulated without incurring additional expense.

In summary, the row spacing that will produce maximum yield is the spacing that produces maximum solar radiation interception (i.e. complete ground cover) by the beginning of reproductive growth. The row spacing meeting this requirement depends on the crop species and plant characteristics. This simple edict is often constrained by technological aspects of the production system, although these constraints are not as important now as they were in earlier days.

Summary

The goal of crop management is, as stated at the beginning of this chapter, to create the perfect environment for the growth of the crop, where the perfect environment is characterized by the absence of stress or other factors that reduce crop growth and yield. This goal may be impossible or uneconomical to achieve, but that does not detract from its usefulness as a goal. The management practices discussed in this chapter are fundamental components of grain production systems that contribute to reaching the goal of the perfect environment. There are many management options available to an individual producer; selecting the best combination is not always easy and it may be constrained by factors outside the realm of the physiological processes controlling crop yield.

The principles underlying the choice of management practices described in this chapter are based on the concepts developed previously in Chapters 2 and 3; they are relatively simple and generally consistent across species and locations. Maximum yield starts with high-quality planting seed that produces adequate plant populations (with uniform spacings and time of emergence for non-plastic species) of a high-yielding variety where the combination of planting date and variety maturity puts the growth cycle in the most favourable environment. The combination of row spacing, plant population, variety maturity and planting date must result in maximum solar radiation interception by the beginning of reproductive growth. While the exact combination of these variables will vary, the combination that meets the objectives outlined here should produce maximum yield for a given environment. There may be situations where the most productive system violates these fundamental rules, but the exception does not negate the value of these rules.

Crop Production in the Future – Challenges and Opportunities

5

> We observe a world of great opportunities disguised as insoluble problems.
> Franklin Delano Roosevelt, US President 1933–1945

Introduction

The grand challenge facing agriculture today is the same one that humans have faced since the beginning of time – obtaining enough food to survive. The specific challenges facing hunter-gatherers were not the same as those faced by a world dependent upon planting and harvesting crops (agriculture), but the overall objective of both systems was the same – acquiring enough food to feed the population.

Thomas Robert Malthus, an influential British economist and clergyman, formalized the challenge of balancing supply and demand in his 1798 publication *An Essay on the Principles of Population* when he wrote:

> I think I may fairly make two postulata. First, that food is necessary to the existence of man. Secondly, that the passion between the sexes is necessary and will remain nearly in its present state. ... I say that the power of population is indefinitely greater than the power in the earth to produce subsistence for man. ... population, when unchecked, increases as a geometrical ratio. Subsistence increases only in an arithmetical ratio. ... By the law of our nature which makes food necessary to the life of man, the effects of these two unequal powers must be kept equal.
>
> (Malthus, 1993, p. 12)

According to Malthus's concepts, population will always grow until restrained by the food supply, at which point many in the population are destitute with barely enough food to survive. Increasing the food supply ultimately increases the number of people who are starving and destitute. Malthus's ideas painted a very dismal picture of the future of humankind.

This apocalyptic view of the future surfaced regularly down through history. Sir William Crooks in 1898 in England, for example, predicted impending

© D.B. Egli 2021. *Applied Crop Physiology: Understanding the Fundamentals of Grain Crop Management* (D.B. Egli)
DOI: 10.1079/9781789245950.0005

famine as the availability of N limited wheat yields (Crooks, 1898). Paul Ehrlich predicted widespread famine in 1970 as population outstripped the food supply (Ehrlich, 1968). Wennblom (1978) suggested that crop yields in the Midwestern USA peaked in 1978. Predictions in 2005 suggested that food production would have to double by 2050 to meet demand (Tilman *et al.*, 2011). This was based on expected population growth and improvements in diets as the gross domestic product (GDP) increased.

Most of these predictions of gloom and doom did not come true. Predictions of future food supplies are often wrong because the predictor could not account for technology that did not exist when the prediction was made. Sir William Crooks' concern with declining soil fertility, for example, was alleviated by the development in the early 1900s of the Haber–Bosch process that fixed N_2 from the air into NH_4 (Hager, 2008). This process provided the N fertilizer that fuelled the Green Revolution and continues to feed the world. It also provided the explosives that allowed Germany to prolong the First World War and enabled the rise of Hitler. We do not know yet if the food supply will have to double by 2050, but there are suggestions that it might not be necessary (Hunter *et al.*, 2017).

The world escaped the apocalypse predicted by Malthus for over 200 years. In other words, Malthus and the other proponents of gloom and doom were spectacularly wrong, but running out of food is a narrative that continues to this day. Predictions of 'doom and gloom' are more newsworthy than optimistic outlooks and they are useful to justify research and increased research funding to avoid the (supposed) coming catastrophe. Gregg Easterbrook presented a more optimistic outlook on the future in his recent book *It's Better than It Looks: Reasons for Optimism in an Age of Fear* (Easterbrook, 2018).

The world population when Malthus published his postulates in 1798 was roughly 1 billion (Cohen, 1995); by mid-2019 it had increased to 7.7 billion (United Nations, 2019), nearly an eightfold increase. The increase over that period was not even, of course, given the way that populations grow; it reached 2.5 billion by 1950 (Cohen, 1995) (an increase of 1.5 billion over 1798) and then increased by 5.2 billion in the roughly 70 years between 1950 and the present. The growth between 1950 and the present coincided with the high-input era of agriculture when yields increased rapidly in many developed countries (see Fig. 1.1 in Chapter 1, this volume). In spite of adding over 5 billion mouths to the table in a relatively short interval, the world population is better fed now than it has ever been – better fed perhaps to its own detriment.

Indications of a well-fed population in today's world are many and varied. For example, society is facing a pandemic of obesity; the rate of obesity in 73 countries doubled between 1980 and 2015 (GBD 2015 Obesity Collaborators, 2017). People's height, which is an indication of the quality of their diet, increased substantially between 1896 and 1996 (Smil, 2019, p. 7). Increases in food production over this interval, a function of increasing area and higher yield, clearly outstripped, on average, the increase in population in spite of the potential differences in the two growth rates. Agriculture had the capacity to create this well-fed world when between one-third and one-half of the food produced was wasted before it was consumed (Foley *et al.*, 2011).

During the high-input era, the excess supply of agricultural commodities in developed countries generally resulted in low grain prices and deployment of numerous government programmes designed to bolster farm income (Schaffer and Ray, 2019). On the other side of the globe, Indian farmers were struggling with huge inventories and low prices for lentil, oilseeds and cereals (Jadhav, 2017). The diversion of food and feed production to the production of bio-fuels from maize (ethanol) and soybean or other oil crops (biodiesel) without creating food shortages as predicted (Cassman and Liska, 2007) is another indicator of the excess capacity of the food production system. An average of 38% of the US maize crop was diverted to ethanol production from 2015 to 2019 (NASS, 2020) and still the price of maize was low. Of course, not all of the maize diverted to ethanol production is lost from food production because the residue left after the production of ethanol (distiller's grains) is used for animal feed, the primary use of maize. The food production system in the USA is the world's largest exporter (Smil, 2019, p. 39) and it also produces feed for 77.8 million dogs, 85.6 million cats (APPA, 2016) and 10 million horses (FAOSTAT, 2020), not an inconsequential demand or an absolute necessity. The dietary energy consumed by dogs and cats is equal to the consumption by about 62 million people (Okin, 2017). In spite of Malthus's arithmetical growth rate, food production in the modern era exceeded population growth, accommodating better diets, excess consumption, and significant non-food uses. However, not all areas of the world can boast of excess production, there are still areas where people do not have enough to eat.

Professor Philip Handler (Bunting, 1991) addressed this apparent contra-diction of plenty and want existing simultaneously in 1978 by formulating three general rules that guide the agricultural economy of the world. The first rule stated that world food production far exceeds the amount necessary to feed the world population. The problem is how to get it from where it is to where it is needed. The second rule suggested that there are no hungry people in the world who have money, i.e. the proximate cause of hunger is a lack of cash to buy food or resources to produce it. The third rule states that no farmers are happy about growing more food than their family can eat, unless some-body gives them something they want in exchange for the extra food. Handler's suggestion that the lack of effective demand restricts the output of food shifts the emphasis of producing more food from the physical/biological world to the economic. If there is demand (i.e. people have money), it will be filled.

Overproduction in developed countries does not eliminate food scarcity in developing countries. In the short run, increasing wheat yields in Kansas pro-vides no benefits to food-scarce countries if they do not have money to buy the wheat. This short-term relationship is at the heart of Handler's rules. In con-trast, the crop scientist asks how increased food demand from an increasing population will be met – higher production must come from a larger area devoted to crop production, from more intensive cropping systems, higher yield or some combination of all three. Handler's rules are best suited to explain the short run (people starving in Africa while an Iowa farmer declares bankruptcy because of low grain prices), while the crop scientist's approach is better suited to considering how to meet future food requirements. Will physical and biological limits intervene and make it impossible to feed the world population in 2050?

Prospects for feeding the world in 2050 depend upon predictions of future supply and demand. When considering predictions it is worthwhile to keep in mind the old Danish proverb often attributed to the famous physicist and Noble Laureate Niels Bohr: 'prediction is very difficult, especially if it's about the future'. Predictions depend upon assumptions (guesses?) which often involve the simple extrapolation of current trends into the future. New technology can, and often does, disrupt the trend and establish a new equilibrium, making the prediction worthless. Examples of these disruptions abound. Sir William Crooks' concerns about diminishing supplies of N fertilizer were completely eliminated by the development, just 15 or so years later, of the Haber–Bosch process to fix N_2 from the atmosphere into NH_4 for N fertilizer (Hager, 2008). Predictions in 2001 suggested that world peak oil production would occur in 2005 (Deffeyes, 2001, p. 158), but technological developments since then, principally the fracking system of oil extraction, boosted production and contributed to current low prices (Yergin, 2020, pp. 14–24). It is impossible to include unknown technology in predictions of future resource availability, making it difficult to accurately predict potential resource shortages.

Predicting food demand requires estimates of future population growth and changes in diet (primarily the amount of meat consumed). The United Nations Department of Economic and Social Affairs forecasts high, median and low population growth rates, illustrating the uncertainty associated with predicting population growth. Its median estimate of the world population in 2050 is 9.7 billion versus 10.56 billion for the high and 8.96 billion for the low estimate (United Nations, 2019). These predictions produce a twofold range (1.26 to 2.86 billion) in the mouths added to the table by 2050. This growth will be unevenly distributed, with just nine countries accounting for over half of the projected median increase by 2050. On the other hand, the population of 55 countries will decline by 1% or more by 2050 (United Nations, 2019). Bricker and Ibbitson (2019, p. 2) predict that world population will peak in about three decades, which is in line with the low prediction by the United Nations (2019). Declining birth rates responsible for static or declining populations are attributed to societal changes resulting in children being a burden instead of an asset (Bricker and Ibbitson, 2019, pp. 46–52). These changes include urbanization (the world was more urban than rural for the first time in 2007), empowerment of women, fewer family interactions and a declining role of religion in peoples' lives.

Population estimates predict a future demand for food that is highly variable, ranging from a modest increase that ends by 2050 (low estimate) to a large increase that is sustained through the end of this century (high estimate).

The trajectory that population follows in the coming decades will, in large part, determine how difficult it will be to match supply and demand for food. The match will be much easier if the projections of Bricker and Ibbitson (2019, pp. 46–52) are accurate. Changes in diet related to the increase in GDP is one of the key assumptions in the predictions of food demand by 2050. Increasing GDP is an indicator of affluence that will result in greater meat consumption requiring more grain production. Estimates that food production needed to double between 2005 and 2050 were based on the median population growth rate and increases in GDP and meat consumption (Hunter *et al.*, 2017).

There is no easy way to compare supply and demand for food over time. The complexity of food sources, international trade that connects food excess areas with deficit areas and variation in diet make it difficult to make simple comparisons. Per capita grain production (kg person^{-1}), however, provides a simple evaluation of changes in sufficiency of grain production. This ratio indicates clearly whether grain production is increasing or decreasing relative to population growth, even though the absolute value of the ratio required to adequately feed the population is not known. The ratio, as calculated here, does not account for non-food uses of the grain, nor does it account for variation in diet. The ratio simply describes the relationship between grain production and population. A stable ratio over time simply suggests that the increase in production is matching the increase in population.

Per capita production of wheat (world) and rice (Asia) increased from 1960 to the early 1980s after which it fluctuated from year to year with no real trend through 2018 (Fig. 5.1), indicating that production kept up with population growth. A stable ratio for 40 years when population in both regions increased by roughly 70% is comforting, but the big question is what will happen in the future. I extrapolated the ratio to 2050 by assuming no change in production area, that population growth followed the median estimate from the United Nations (2019) and three scenarios for yield growth. Extrapolating the historical linear increase in yield from 2009 to 2018 to 2050 resulted in a steady increase in per capita production for both crops (Fig. 5.1). Maintaining the linear increase in yield, with no change in cropping area, was all that was needed to increase production per capita for both crops (20% by 2050) with the median population increase. Surprisingly, the ratio could be maintained at its recent level through 2050 with one-half of the linear yield growth rate (Fig. 5.1). An increase in the yield growth rate over the recent historical rate was not needed to maintain the status quo; in fact, a 50% reduction could be tolerated with no change in the ratio. This analysis suggests that the problem of feeding the world in 2050 may not be as difficult as many suggest. Some yield growth was, however, required; the per capita availability in 2050 was reduced (12 to 20%) when the ratio was calculated assuming no increase in production (constant area and yield remaining at the 2020 level, estimated by extrapolating the 2009 to 2018 linear increase to 2020) through 2050. Handler's conclusion (Bunting, 1991) that world food production exceeds the needs of the world population applies only to the present; increasing population

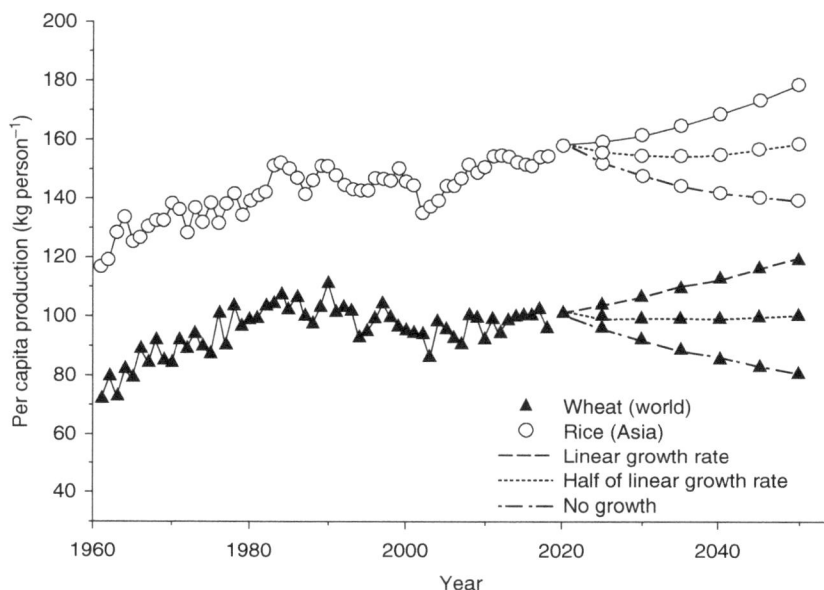

Fig. 5.1. Per capita production of rice (Asia only) and wheat (world) from 1961 through 2018. Production and population data from FAOSTAT (2020). Extrapolation of the ratio from 2020 to 2050 was based on the median population growth rate (United Nations, 2019), no change in the production area and yield growth that was maintained at the linear growth rate from 2009 to 2018, one-half of the linear growth rate or no yield growth.

requires an increase in production or the per capita supply will decline (Fig. 5.1). The analysis presented in Fig. 5.1 does not consider the increases in grain production needed if per capita meat consumption increases.

The assumed rate of population growth has a direct effect on the projections of per capita production in Fig. 5.1. Maintaining the per capita level for wheat through 2050 required a 36% increase in production at the high population growth rate versus only 14% at the low rate (the one-half yield growth rate with a constant area increased total production by 24%). Rice required an increase of 21% (high population growth rate) or 1% (low rate) in total production to maintain the 2020 ratio through 2050. These calculations illustrate the importance of assumptions when projecting food sufficiency into the future.

Population growth rates are declining in many countries; in fact, the United Nations Department of Economic and Social Affairs estimates that 55 countries will lose population by 2050 (United Nations, 2019). Considering recent social trends it seems that future growth rates may approximate the low rate, vastly simplifying the problem of producing a well-fed world by 2050. Uneven global distribution of demand (population and quality of diet) and production resources (soil, water, climate) are certainly complicating issues, as is the potential disruption of climate change. The trends in Fig. 5.1 provide some indication of future expectations that are not as grim as often portrayed by the gloom and doomers.

There are many avenues for increasing food production in the next 30 years. Increasing yield is not the only route to a well-fed world; there are other opportunities using more or less proven technology. Modest increases in the area in production without damaging the environment may be possible at some locations. Increasing the cropping intensity (more crops per year) effectively increases the area devoted to crop production without clearing forests or bringing new land into production. Interestingly, the warming climate and longer growing seasons in temperate climates will increase opportunities for multiple cropping, such as growing a summer grain crop after a winter grain crop (i.e. the wheat–soybean double-cropping system). There are several opportunities to use the food that is produced more efficiently (i.e. feed more people from a given level of production). As much as 30 to 50% of the food produced is wasted (Foley *et al.*, 2011); reducing waste would be the same as increasing production. Changing diets to reduce meat consumption (especially beef) and decreasing non-food uses, including biofuels and feeding the huge army of cats, dogs and pleasure horses in the USA, would increase food availability.

There are a number of opportunities to apply new technologies that may help feed the world's population in 2050. Plant breeding and the development of new, improved varieties played a significant role in historical increases in yield. The massive improvements in our ability to manipulate plants genetically over the past few decades could increase that contribution in the future. A whole host of new technologies that will change the way we manage grain crops are available or under development. Precision agriculture techniques, remote sensing by drones or satellites, and new techniques to summarize, evaluate and interpret data are rapidly coming online. It is possible that new crop species will play a role in increasing food production in the next 30 years. There is growing interest in new forms of agriculture (organic agriculture, agroecological approaches, regenerative agriculture, producing 'meat' from plants or in culture systems) that are touted as the key to feeding the world in a sustainable manner. These options will be discussed in the following pages.

Increasing food production in the next 30 years will, no doubt, face new challenges that were not issues in the past. The biggest challenge to increasing production will be the changing climate resulting from increases in the concentration of greenhouse gases (CO_2, CH_4 and N_2O) in the atmosphere. Closely linked to climate change is the availability of energy. Modern high-input agricultural systems function most efficiently when adequate supplies of cheap energy are available. Energy cost and availability as society switches from carbon-based energy systems to other sources remain to be determined.

Other challenges facing agriculture in the next 30 years include declining resource availability. The fixed supplies of some fertilizer materials may limit crop production in the future, while supplies of high-quality water for irrigation are already limiting at some locations. Water availability, of course, is closely related to climate change. It is very difficult to predict future resource availability and its impact on food production systems, given the role that technology plays in its availability.

Malthus and all of the other gloom and doom purveyors down through history were wrong; the world successfully beat the Malthusian trap for 220 years. Can the food supply increase enough to maintain the per capita production for the next 30 years? When one considers the historical successes at increasing yields and production, the new technologies available to increase yield and the numerous options available beyond increasing yield, it is easy to take an optimistic view of our chances of feeding 9.7 billion people (or will it be only 8.9 billion?) by 2050. We will discuss some of these options in more detail in the remainder of this chapter, but the wild card in this discussion is the effect of climate change on agricultural production and our ability to adjust to the change. As we shall see, the potential disruption of the system could be catastrophic, and the general public seems to show little interest in reducing emissions of greenhouse gases to minimize climate change. Unfortunately, when climate change reduces the food in your local grocery store, it will probably be too late to intervene effectively.

Climate Change

Crop productivity is determined in large part by the environment the crop is growing in; both the above- and below-ground environment are important, but producers probably worry more about the above-ground environment. After all, what are producers obsessed with all summer? The weather (the above-ground environment). Climate can be defined as the 'average state of the weather at a particular location', or as Mark Twain, the great American author and humourist, described it, 'climate is what you expect, weather is what you get'. Weather is what is happening today – it's raining or it's not raining; climate refers to the average rainfall on this date. Crop management practices are adapted to the climate at a given location, following the logic that past climates are a prediction of future climates. This relationship does not hold if the climate is changing.

Does climate, the average state of the weather, change? When glaciers covered parts of the northern hemisphere during the ice ages, the climate was definitely different, a change that persisted for thousands of years. Climate can also change on much shorter time scales, although some climatologists argue that short-term changes are just fluctuations, after which the climate returns to its 'normal' state. Change versus fluctuation is very much a matter of perception. Humans perceive a fluctuation that lasts for several hundred years as a change. The ice ages appear as a fluctuation on a geological time scale covering millions of years. The multi-year drought that caused the dust bowl in Midwestern USA in the 1930s is an example of short-term climate change; 10 years after it started, it was over (Egan, 2006). An even shorter-term change occurred after the 1815 eruption of the Mt Tambora volcano in the Indonesian archipelago. The volcano sent a plume of ash 29 km (18 miles) into the sky and that ash cloud spread around the globe and lowered temperatures worldwide, causing the 'year without summer' in 1816. Freezing temperatures and snowstorms occurred in June across Canada, the northern USA and Europe, causing widespread crop failures, food shortages and social disruption (Klingaman and Klingaman, 2013).

The mechanisms responsible for climate change are, in some cases, known, but in others they are not. The ice ages were caused by variation in the characteristics of the earth's orbit around the sun that reduced the solar radiation reaching the earth's surface, lowering temperatures and triggering the growth of glaciers. Volcanos put huge amounts of debris and aerosols into the stratosphere where it blocks solar radiation and cools the surface. The earth is currently experiencing another change in climate – an increase in temperature – that is driven by an enhanced greenhouse effect resulting primarily from humankind's combustion of fossil fuels (coal, petroleum, natural gas). This change will be long-lived, lasting until fossil fuels are phased out and the concentration of greenhouse gases in the atmosphere decreases.

The surface of the earth is warmed by absorption of the sun's rays. The earth's surface, following the laws of physics, cools by emitting radiation. Radiation from the sun is short-wave radiation because the sun has a very high temperature (5799 K (9979°F)); the surface of the earth is much cooler (289 K (61°F)), so the radiation emitted by the earth is long-wave radiation that is invisible to the human eye. The difference in wavelength is important because only long-wave radiation is absorbed by gases in the atmosphere (greenhouse gases: CO_2, CH_4, N_2O and water vapour) and reradiated back to the earth's surface. The absorption of long-wave radiation by these gases and the reradiation back to the earth's surface reduces the amount of cooling and increases the surface temperature of the earth. This phenomenon is called the greenhouse effect because it is similar to what happens inside a greenhouse. Solar radiation passes through the glass roof, but the long-wave radiation from the floor, benches and plants is absorbed by the glass and the temperature inside the greenhouse increases. The greenhouse gases act like a blanket to warm the earth and the higher the concentration of these gases in the atmosphere, the thicker the 'blanket' and the more warming there is.

The existence of the greenhouse effect was first proposed by Joseph Fourier in 1824. Svante Arrhenius predicted in 1896, more than 120 years ago, that CO_2 released by burning coal would eventually cause surface temperatures to rise. In fact, he laboriously calculated that doubling the CO_2 concentration would increase global temperatures by 6°C (10.8°F), a rise that is similar to current predictions by sophisticated models (Rennie, 2020). His prediction was largely ignored, but now we see it coming true – there is no doubt that air temperatures are rising.

There are several examples of the greenhouse effect in nature. Desert climates are known for their very hot days and relatively cold nights, caused by the reduced greenhouse effect resulting from the low levels of water vapour in the air. The coldest winter nights in temperate climates often occur when there is no cloud cover; clouds are a mixture of water droplets and ice crystals that enhance the greenhouse effect. The difference between day and night temperatures on planets without an atmosphere and therefore no greenhouse effect is much larger than it is on earth. Clearly, the greenhouse effect is real; the debate about climate change centres more on the effect of warming the earth and atmosphere on climate than it does on the existence of the greenhouse effect per se.

The concentration of CO_2 in the atmosphere increased steadily from 280 ppm at the dawn of the Industrial Revolution to 412 ppm today. CH_4 and N_2O concentrations are also increasing, but they are present in the atmosphere in much lower concentrations (parts per billion) than CO_2. Water vapour is not involved in global warming because, on the average, its concentration in the atmosphere is not increasing. The primary source of CO_2 is the combustion of fossil fuels, while agricultural activities are important sources of CH_4 (rice paddies, ruminant animals) and N_2O (fertilizer). The combustion of fossil fuels by the agriculture sector also contributes to the greenhouse effect. Some experts estimate that as much as 30% of greenhouse gas emissions come from the global food system (Clark *et al.*, 2020).

Climate change and global warming were hotly debated for many years. Were we simply experiencing normal fluctuations in the weather or were we observing a change in climate? Most of these arguments were about the effect of increasing temperatures on our weather and climate. Are wet years or droughts or weather disasters a result of climate change? Will there be more or fewer hurricanes? Will we have longer and more severe droughts or will it be wetter with more high-intensity rainfall events? All of these issues were and are hotly debated, but I am not aware of anyone debating the fundamental cause of climate change: the absorption of long-wave radiation by greenhouse gases and reradiating it back to the surface, causing an increase in temperature. No one argues that the greenhouse effect does not exist. Accepting that, we know the earth will get warmer, we just do not know exactly what localized changes in weather will occur.

How much has the climate changed? The global temperature has increased by about 1°C (1.8°F) over the 1950–1981 mean (the increase is much larger in polar areas). Experts predict that increasing the CO_2 concentration in the atmosphere to 600 ppm will cause an increase of 1 to 4°C (1.8 to 7.2°F). Identifying the effect of the increased temperature on our weather is difficult because the change is small relative to normal day-to-day variation. Normal variation in temperature from day to night is 11.1 to 16.7°C (20 to 30°F); some days, weeks and months are hotter than normal, others are cooler than normal; summer is much warmer than winter – this variation is large relative to the small changes due to the enhanced greenhouse effect, making it difficult to document or personally experience climate change. This difficulty makes it easier to debate whether climate change exists or not. Identifying the change is one issue, but a more important issue is predicting what changes will occur in the future.

Predicting the effect of increasing greenhouse gas concentrations on climate is difficult, especially when trying to predict changes at specific locations, not just broad generalities. We are used to running experiments to determine how crops respond to changes in their environment, but we cannot run experiments to estimate the effect of increasing greenhouse gas concentrations on climate. Consequently, scientists are forced to use models of weather and climate to study global warming and predict future climates. General Circulation Models (GCMs) use the fundamental laws of physics to predict the circulation patterns in the atmosphere that give rise to weather and climate at the surface. These models are similar to the models that produce the forecast you see on the Weather Channel or the nightly news, except that they are focused on long-term changes in weather and climate (10, 20 or 50 years in the future) whereas the model producing tomorrow's forecast only has to predict conditions for several days. All models, whether they are GCMs or models that predict crop growth and yield, represent simplifications of very complex systems. Models are based on what is known about the system being modelled; they cannot include relationships that are not known. For these reasons, models are far from perfect; in fact, there is an adage among modellers that 'all models are wrong, but some are useful'.

Global circulation models are huge, horrendously complex computer programs that require very fast supercomputers to produce results in a reasonable amount of time. All GCMs do not produce the same results when given the same initial information, which is not surprising since they represent slightly different simplifications of the 'real' system. These models are the only tool we have to predict the effect of increasing greenhouse gas concentrations on future climatic conditions. The output from GCMs is often used as input for crop simulation models to predict the effect of climate change on future crop yields. Using the output from a suite of GCMs as input to a suite of crop simulation models makes it possible to look for common responses, which, because they are common to a group of GCMs and crop simulation models, have a greater probability of being correct.

What effect will climate change have on grain crop production? Crops growing in the new climate will likely be exposed to higher concentrations of CO_2, higher temperatures and possibly changes in water availability. A summary of data from the literature suggested that increasing temperature by 0.8°C (1.4°F) reduced irrigated yield of maize, wheat, rice, sorghum, groundnut and bean growing in their area of adaptation in the USA by 2.5 to 8% (Hatfield *et al.*, 2011). Soybean yield in the US Midwest region increased (+1.7%), however, because the background temperature was below optimum, so increasing temperature increased photosynthesis and yield. Temperatures in southern US soybean-growing regions were higher, so a rise in temperature decreased soybean yield (–2.4%) (Hatfield *et al.*, 2011). Assuming that the effect of rising temperatures on growth and yield is always negative is not correct; it depends on the background temperature. In climates with lower temperatures, warming may increase yield.

When a higher CO_2 concentration (440 ppm, current level is 414 ppm) was included in the evaluation to better simulate real-world conditions, the yield response of most C_3 crops switched from negative to positive. The stimulating effect of high CO_2 on photosynthesis (see Chapter 2, this volume) overcame the negative effects of high temperature. The yield of maize and sorghum, C_4 crops whose photosynthesis does not benefit from high CO_2 levels, increased by only 1%. The analysis of Hatfield *et al.* (2011) did not include effects of temperature extremes. Extremely high temperature events that disrupt pollination of maize and other crops are expected to increase as the earth warms and these disruptions could cause catastrophic yield loss.

The effects of water availability on yield were not considered in the previous discussion, but excess or deficit rainfall could also have major effects on future crop productivity. Climate change is expected to affect both extremes – too much water and not enough water – and both could cause significant reductions in yield. Recent droughts in California and the resulting depletion of the groundwater resources for irrigation are probably related to climate change. A warmer atmosphere can hold more water vapour which could contribute to higher-intensity rainfall events, damaging crops and reducing yields. Higher temperatures will also increase ET rates (Hatfield *et al.*, 2011). The availability of water will be, in my opinion, a key aspect of climate change and it is more complex than temperature changes because the distribution of rainfall, relative to crop development, is so important. Timing of a dry spell determines whether or not it will affect yield. Global climate models cannot predict these localized effects of climate change, making it hard to evaluate their potential effects on yield. It is clear, however, that catastrophic yield losses could occur in many locations if climate change creates a drier environment (less rainfall) or more frequent extreme events (too much rainfall or drought).

It may be possible to minimize some of the negative effects of climate change on grain yields by modifying management practices. Obviously, drought can be eliminated by irrigation, if water is available, but extended drought conditions could reduce the availability of irrigation water. Higher temperatures cause the spring warm-up to occur earlier in temperate climates, resulting in a longer growing season (Linderholm, 2006). Early planting of summer annual grain crops coupled with the use of short-season varieties might allow the crops to avoid some of the high-temperature stress in mid-summer. Longer growing seasons will increase the opportunities for double cropping a summer grain crop after a winter crop to potentially increase total productivity per year. Global warming may make it possible to grow grain crops in more northerly locations that will have adequate growing seasons and, hopefully, somewhat cooler temperatures, thereby avoiding high-temperature stress. The recent increase in maize production in the prairie provinces of Canada was facilitated by longer growing seasons in these more northerly locations (Bjerga, 2012).

Unabated global climate change could cause catastrophic reductions of crop yields in the long run, but as just noted, it is not entirely clear what the effect will be in the next 30 years (by 2050). Some expert observers suggest that climate change will have minimal effects on crop yield by 2030 (Fischer and Conner, 2018). Others predict significant reductions in yield. Precise predictions of the effect of increasing greenhouse gas concentrations on local climates would be very helpful in preparing for the future, but such predictions are not yet available. Responding to changes in climate as they occur may be the only route available, but it is not as effective, given the time required to develop stress-tolerant varieties and research new crops or new management practices for crops at new locations. This lag time could result in significant reductions in the rate of growth of yield and food production during the adjustment period. Plant breeding programmes test their progeny in the environment where they will be grown, which should result in a gradual adaptation to the 'new' environments. It is not yet clear if this process will keep up with the changing climate or if there will be significant yield loss. As noted in the introduction to this chapter, food sufficiency is determined by supply (production) and demand (determined in large part by population). If, as some suggest (Bricker and Ibbitson, 2019), the world population starts declining in three decades (i.e. by 2050), maintaining adequate food production in the face of climate change will be much easier.

In spite of the clear evidence that the climate is changing, society has historically shown a great reluctance to take the necessary steps to reduce CO_2 emissions and stabilize climate. Reducing greenhouse gas emissions will not reduce the CO_2 concentration in the atmosphere; it will only stop the increase. The CO_2 concentration will have to be reduced to reverse climate change. Planting trees, increasing soil organic matter levels, or physically removing CO_2 and storing it deep in the earth are some of the schemes proposed to accomplish this reduction. The current development of wind and solar power and the interest in electric cars suggest that the general populace and governments are starting to take climate change seriously. It may, however, take rising food costs, starving people and the regular occurrence of devastating storms causing widespread loss of life and property to goad society into making the sacrifices necessary to limit and reverse climate change. We know what to do, but it will not happen until the public wants it done.

Molecular Biology, GMOs and Variety Improvement

The increase in crop yield that started at the beginning of the high-input era of modern agriculture was driven in large part by the development of new improved varieties by plant breeders (see 'Variety Selection' section in Chapter 4, this volume). The development of techniques that made it possible to transfer single genes from one plant to another (within and between species) in the late 1980s was widely touted as a technique that would revolutionize agriculture and traditional plant breeding. No longer would breeders be limited by the genetic variation within a species; now the entire plant kingdom would be at their beck and call. Enthusiasts made extravagant claims that they could easily alter stress tolerance, seed composition and yield. There was no limit to what could be done by genetic engineering. Many expected this exciting breakthrough to solve the problems limiting the efforts of traditional plant breeders and open a new golden era of variety improvement.

Traditional plant breeding, dating from the early 1900s, involved manually transferring pollen from the male to the female parent and, after selfing for several generations, selecting offspring that had desirable traits (higher yield, lodging, shattering or disease resistance, etc.). These selections were compared with current varieties and, if they were superior, they were released as new varieties. The dramatic increase in yield in the high-input era of grain crop production (see Fig. 1.1, Chapter 1, this volume) provides convincing evidence that this breeding approach was highly successful, in spite of some serious limitations. First, it is slow; it often takes 8 to 10 years from the first cross until a variety is ready for release to producers. Second, plant breeders could only work with the genetic variation naturally available within the species of interest or they could try to create useful variation via mutagenesis, a difficult proposition. It was hoped that genetic engineering would solve these problems.

Traditional plant breeders sped up the process by establishing winter nurseries in warm climates or in greenhouses to produce several generations in a single year, but the final yield testing had to be done in the environment where the variety would be grown. C.M. Donald, an Australian plant breeder and crop physiologist, pioneered the idea that selecting for yield would be more successful if the breeder knew what plant characteristics were associated with high yield and selected for those characteristics. Donald developed the crop ideotype concept (Donald, 1968), where the ideotype represented a description of the characteristics of a high-yielding variety that would guide the plant breeder's choice of parents and offspring and make the breeding process more efficient. The ideotype concept triggered the golden age of Crop Physiology as the hunt was on for traits that contributed to high yield; traits that could be used to develop ideotypes. Unfortunately, the ideotype approach was not very successful and plant breeders rarely used it.

The ideotype approach failed because some of the traits recommended by crop physiologists were not intrinsically related to yield. The trait was often simply correlated with yield in a set of varieties; rarely was there any evidence of cause and effect. Complex, hard-to-measure traits failed because they could not be accurately measured on the large number of plants in a breeding programme. Plant breeders' reluctance to devote time and effort to unproven techniques also contributed to the failure of the ideotype approach. Finally, it is possible that the concept that selecting for a single trait or a group of traits associated with yield is a shorter route to high yield is just wrong. The production of yield is a complex, many-faceted process; too complex perhaps to be captured by one or two traits. Perhaps yield can only be increased by selecting for yield. The end result of the ideotype approach was that plant breeders ignored the suggestions of crop physiologists and continued simply selecting for yield.

Genetic engineering was supposed to eliminate many of the problems faced by traditional plant breeders and speed up the production of improved varieties. Today, however, roughly 40 years after genetic engineering burst on to the scene with its extravagant promises, the only products of this technology in the grain farmer's field are varieties with herbicide tolerance, resistance to some insects conferred by the *Bacillus thuringiensis* (Bt) protein and a few varieties with drought tolerance. There are apparently a few varieties modified directly to produce higher yields in the pipeline. A remarkably minuscule showing after all the initial hype. Herbicide tolerance and Bt had a major impact on grain crop management. Tolerance to the herbicide Roundup (glyphosate) made weed control in maize and soybean much easier and much cheaper; no longer did the producer have to mix and match herbicides with the weed spectrum in each field. If weed control was easier, the odds are it was better, and better weed control would increase yield. The same logic can be supplied to the use of Bt varieties. Now, 20 some years after their introduction, the development of herbicide-tolerant weeds and insects that are resistant to Bt are limiting the usefulness of this technology. Glyphosate-resistant weeds are countered by the development of varieties that are tolerant of other herbicides. It is not clear if this is a solution to the problem or simply a rapidly spinning resistant weed–herbicide tolerance treadmill.

After all the promises and hype, why has this technology had such a limited effect on yield potential? The scientific literature is full of papers identifying genes that are related to yield (e.g. reviews by Van Camp, 2005; Zhang, 2007; Dunwell, 2010) and they continue to appear (Wu *et al.*, 2019; Zhang *et al.*, 2019). Why is it so difficult to translate these yield-controlling genes into varieties with higher yield? Part of the problem, in my opinion, stems from the way molecular biologists evaluate their modified plants. They often evaluate yield with isolated plants in greenhouses, not in plant communities. Traits that confer higher yield on isolated plants do not necessarily increase yield of the plant communities that make up producers' fields. Increasing plant height or tillering or branching ability, for example, will probably increase yield of isolated plants, but will have no effect on plants growing in a community. Molecular biologists often evaluate drought tolerance as the ability of the plant to survive drought, but survival is not an issue in the producer's field. The successful drought-tolerant variety must show a smaller yield reduction under stress compared with a less tolerant variety and, ideally, produce yield equal to the best variety in the absence of stress. Many of the yield-controlling genes probably failed to produce commercial varieties because their relationship to yield did not translate to field communities (i.e. the producer's field).

Expectations that were too great may have contributed to the apparent failure of genetic engineering for yield. The production of yield is a complex, many-faceted process (see Chapters 2 and 3, this volume), so it was probably unrealistic to think that a single gene would control it. Genetic engineering for higher yield and the crop ideotype approach of selecting for individual plant traits utilized a similar approach to increasing yield. Both used individual traits thought to make the plant more productive and both, by and large, have not been very successful. Perhaps the scientific community should have realized sooner that these approaches were unlikely to be more successful than selecting for the end product – yield.

Molecular biology techniques may be more useful as a complement to standard breeding techniques for the production of improved varieties. Molecular markers, genomic selection and other approaches are being integrated into many plant breeding programmes to speed up the production of new varieties. These approaches are a far cry from the original dream of inserting a gene to increase yield or make plants tolerant of stress, but they are effective none the less, and, interestingly, they ultimately rely on selection for yield. Shortening the time from the first cross to the release of a new variety improves the rate of yield gain and that will accelerate increases in the food supply.

The genetic revolution has not yet delivered on its original promise. The technology it spawned is amazing when compared with that available to the original plant breeders, and it continues to evolve, so I think it is a mistake to discount its potential contributions. These amazing techniques will no doubt help feed the world's population by 2050.

Precision Agriculture/Big Data

Will the application of advanced technology to managing grain production increase yield and efficiency? A wide range of technologies, from sensors on satellites or on field equipment to gridded soil sampling and computer algorithms, are being brought to bear on management issues. These approaches are grouped under the general heading of precision agriculture, expressing the idea that management will be applied at a finer scale than ever before. No longer will the field be the management unit; now management will be localized to take advantage of sub-field variation in soil and plant characteristics.

Precision agriculture is based on the concept of applying the right amounts of inputs in the right place at the right time (Thompson *et al.*, 2019). It involves collecting much more information about the crop and its environment (primarily the soil environment) and using it to adjust inputs and make management decisions. The highly touted potential of many aspects of precision agriculture has been slow to be realized by producers, so, in this respect, it is somewhat similar to genetic engineering.

The development of some precision agriculture techniques was a matter of technology looking for a use, not an identified problem looking for a technological solution. The technologists said we can do 'this', but it was often difficult to find an aspect of grain crop management that would benefit from 'this'.

Finding a useful aspect would be easier if grain crops required more management decisions in real time. Some of the precision agriculture technology is truly amazing; surely, it will be possible to develop useful applications for it, useful in the sense that the producer sees the benefit to his bottom line and adopts the technology.

The global positioning system (GPS) made precision agriculture possible. Locations in a field can be identified very accurately (up to ~2.5 cm (1.0 in)) with GPS, so equipment can find a location on subsequent passes. There would be no precision agriculture without GPS.

A wide variety of information is collected under the umbrella of precision agriculture. Yield monitors on combines describe variation of yield in a field, while gridded soil sampling characterizes variation in soil fertility levels. A variety of sensors on equipment (e.g. planters, liquid fertilizer applicators), drones and satellites measure various aspects of the soil and the crop, providing a blizzard of information, much of it in real time. Some of these sensors seem like magic, when, for example, satellites measure photosynthesis, crop water use (ET) and soil moisture from thousands of kilometres above the earth's surface. Equipment currently available includes guidance and auto-steer systems, yield monitors and variable-rate applicators of fertilizer, seed, pesticides and irrigation water, to mention just a few.

The challenge facing precision agriculture is, and always has been, to find something useful to do with it. Useful meaning something that puts more money in producers' pockets, either from higher yield or from the same yield with fewer inputs (greater efficiency) or some combination thereof. Reducing environmental damage (e.g. avoiding excess fertilization) is also a possible benefit from the application of precision agriculture techniques that doesn't fit with making or saving money, at least in the short run. In my opinion, improving efficiency seems to be more likely than increasing yield. Reducing fertilizer rates or seeding rates on lower-yielding areas of a field can reduce input costs without affecting yield, a clear increase in efficiency. Auto-steer systems, sprayer boom control and planter-row shutoff systems result in more precise placement of pesticides, herbicides and seed, thereby reducing inputs and increasing efficiency. Using a yield monitor to identify low-yielding areas of a field that do not show a profit and removing them from production would increase the overall profitability of that field. Yields will be increased only if historical whole-field management results in areas with less than optimum levels of inputs (e.g. under-fertilized or maize seeding rates that are too low). If farmers are managing the field for its highest-yielding area, greater efficiency is much more likely than higher yield.

The success of precision agriculture techniques depends, first, on variation within a field; without variation, there is no need for precision agriculture. It is interesting to speculate that in-field variation might be less in higher-yielding (uniform soils) and greater in low-yielding locales (variable soil conditions). Utilizing the variation depends upon understanding the relationship between the input and the character that is varying. Reducing seeding rate of maize on low-yielding areas of a field, for example, requires an understanding of the yield–population relationship as well as ear and kernel size characteristics of individual hybrids (Egli, 2019) to estimate the proper population. Variable-rate fertilizer application requires an understanding of the fertilizer rate–soil test relationships. Matching variety to soil type requires knowledge of performance of a variety on various soil types. Defining these relationships for each variety takes time that is not always available given the short shelf life of many varieties; consequently, the information may not be available.

Another aspect of precision agriculture is the advent of 'big data'. Data characterizing the environment (above and below ground), growth and productivity of a crop are accumulated and manipulated by proprietary algorithms, artificial intelligence and machine learning to predict best management practices. Commercial companies offer these services to producers and the information often flows directly from sensors on the producer's field equipment to the companies' computers, requiring very little effort by the producer. The big question is, do these programmes put money in the producer's pocket? A second, and more important, question is, how do we answer the first question? The recommendations, generated by faceless computer algorithms, are specific for a field or a location in a field, making it difficult to determine the value of the recommendation with traditional experimentation. Accepting, without question, the recommendation of a computer algorithm is not very satisfying.

In summary, the technology that makes up all aspects of precision agriculture and the information that can be gleaned from it is impressive, bordering on magic in some cases. The challenge of finding a profitable use for the technology has delayed adoption by producers. A recent survey of grain crop producers in the USA, however, found that a substantial percentage of large grain producers (farming > 405 ha (>1000 acres)) are using some aspect of precision agriculture (e.g. 93% are using yield monitors and 73% are using variable-rate fertilization) (Thompson *et al.*, 2019). It is difficult to predict the future trajectory and contribution of precision agriculture, but it would be a mistake to discount completely future contributions of such an impressive technology. At present, it seems that increasing efficiency and avoiding environmental damage will probably be the primary contribution of precision agriculture to increasing future food supplies. These contributions are not as flashy as increasing yield, but they are solid contributions, none the less.

New Crops

Will new crops be the key to withstanding the ravages of climate change and feeding the 9.7 billion in 2050? There are those who think that new crops, i.e. plant species not widely used now, are the key to solving this problem. Historically, grain crops provided two-thirds of the calories and half of the protein in our diet (Vaughan and Geissler, 1997, p. xvi). Currently, only four grain crops – maize, wheat, rice and soybean – dominate world production, producing 85% of the total production of the 20 grain crops listed in Table 1.1. The potential number of species that could be used for food production, however, is huge. The late Jack Harlan, a renowned expert on crop domestication at the University of Illinois, compiled a short list of 352 plant species that were used for food (Harlan, 1992). Over the ages, humankind narrowed that list down to the point where four species dominate our food supply.

Why did humankind settle on such a small group of crop species? Why did domestication overlook the species now being considered as new crops? Perhaps the crops grown today were best suited for domestication (Sinclair and Sinclair, 2010, pp. 15–23) or perhaps they were domesticated first and then maintained because no one wanted to start domestication over (Warren, 2015, pp. 164–167).

Creating a new crop to challenge the established big four may seem foolhardy, but it has been done. Soybean, first grown for grain in the USA in the early 1900s, is now a member of the elite four. Canola, a crop created when, in the 1960s, plant breeders in Canada removed toxic compounds from oil (erucic acid) and meal (glucosinolates) in seeds of rape, is widely grown today (see Table 1.1). Successfully creating a new crop is possible.

Species currently under consideration as new crops include grain amaranth (Gelinas and Seguin, 2008), chia (Jamboonsri *et al.*, 2012), quinoa, hemp (Pszczola, 2012), vernonia (Shimelis *et al.*, 2008) and potato bean, a legume that produces edible tubers (Belamkar *et al.*, 2015). Attempts are also under way to develop perennial grain crops. The stability and productivity of the prairie ecosystem in the Midwestern USA provided the stimulus to develop perennial grain crops (Jackson, 1985). Perennial crops would provide continuous ground cover, deeper roots to access more water and nutrients in the soil profile and a longer life cycle that may translate into higher grain yields (Glover *et al.*, 2010; Kantar *et al.*, 2016). A longer life cycle, however, will translate into higher yield only if it is associated with a longer seed-filling period (as discussed in Chapter 3, this volume); unfortunately, a longer life cycle does not guarantee a longer seed-filling period. Progress has been slow because the parent species were often low yielding, but a variety ('Kernza') developed from intermediate wheat grass by the Land Institute and the universities of Manitoba and Minnesota is currently in the early stages of commercialization (Zabinski, 2020, p. 182). Enthusiasts rarely discuss the longevity of a productive stand of these perennial grain crops, but that is a key aspect of the system. Much of the value of perennial grain crops will be lost if they have to be reseeded every couple of years to maintain productivity.

Cheng (2018) recently suggested focusing on underutilized crops known for their heat and drought tolerance to maintain food supplies in the face of climate change. The crops he recommended (grain amaranth, teff, millets, buckwheat and quinoa) are often grown in high-stress environments. Improving underutilized crop species should be easier than starting with undomesticated species.

Proposed new crop species are almost invariably described as producing seeds that are highly nutritious on plants that require very little fertilizer or water. These characteristics make them ideally suited for infertile, unproductive soils where they would not compete with established crops. This reputation may not be entirely deserved, because all plants require adequate supplies of mineral nutrients and water to be productive. Perhaps their inherently low yield limits their response to soil nutrients and water, giving the impression that they do not require fertile soils.

In spite of the record of soybean and canola, developing a successful new crop is difficult. A successful new crop must supply something that society needs or wants. Since the crops currently grown satisfy humankind's needs, a new crop may have to compete directly with them on an economic basis, making yield and the cost of production critical. If the new crop is the only supplier in a unique niche market (e.g. the current burgeoning market for cannabidiol (CBD) from hemp plants), it will not have to compete directly with established crops. The key attribute of many new crops under consideration is a more nutritious seed, but whether a successful market niche can be created on that basis remains to be determined.

A new crop species is typically low yielding and often has many undesirable characteristics (lodging susceptibility, shattering, lack of disease resistance, etc.), characteristics that have been bred out of the grain crops currently in production. Breeding will slowly improve the new crop and increase its yield, but at the same time breeders will increase yield of the established crops, so the new crop may never catch up. It may always yield less. Osterberg *et al.* (2017) recently proposed that genome editing might be useful to speed up the domestication process and reduce the time it takes to produce new crops that are competitive with established crops.

It seems unlikely that new crops will contribute significantly to increasing food supplies in the 30 years until 2050; most of them, if they are successful, will probably be relegated to niche markets based on their nutritional value or other unique characteristics. The exception to this somewhat dim outlook will be new crops that flourish because their stress tolerance makes them an essential tool for dealing with runaway climate change.

New Approaches to Agriculture

Are completely new approaches to agriculture needed to ensure adequate supplies of food in the future? Enthusiasts argue that food production systems that are sustainable and environmentally friendly are needed to replace current high-input 'unsustainable' systems. The lack of sustainability of current systems is often attributed to the large quantity of off-farm inputs they require, including synthetic fertilizers that 'poison' the soil, herbicides, pesticides and substantial amounts of energy from petroleum and natural gas. Development of pests and weeds with resistance to pesticides and herbicides, destruction of biodiversity and negative on- and off-site effects on the environment are also taken as indicators of unsustainability. The proposed new approaches would greatly reduce or eliminate many of these problems, but would they do so at the cost of reducing the food supply and making food more expensive?

The new approaches have their roots in the organic agriculture movement that began in the 1840s as an alternative to so-called industrial agriculture (Conford and Dimbleby, 2001, p. 17). Organic agriculture was based on closed nutrient cycles (no off-farm inputs), so it banned the use of synthetic fertilizers, pesticides and herbicides and promoted crop rotations. Organic agriculture was not just a matter of rejecting technology; proponents accepted the natural order of life and intended to work within this order. The early promoters of organic agriculture were definitely anti-science and technology and approached agriculture from a religious viewpoint. They did not believe in experimentation, the very heart of modern high-input agriculture, but felt that the 'farmer just knew what worked'. Successful farming was possible only when the farmer had a spiritual attitude toward the land (Conford and Dimbleby, 2001, pp. 74 and 80).

Other new approaches that are related to organic agriculture include agroecological farming based on ecological principles; regenerative agriculture that embraces organic principles to replenish and strengthen the soil; and permaculture that favours the use of perennials, polyculture and zone designs based on landscape characteristics. New approaches that are not closely tied to the organic agriculture movement include the local food movement that emphasizes the reduction in 'food miles' and transportation costs (both in the price of the food and reducing the CO_2 emissions associated with transport); 'vertical' farming where food is produced in controlled environments that supply radiation, water, nutrients and appropriate temperatures to maximize productivity (Despommier, 2010); and the production of 'meat' from plant products or grown from animal cells in culture – the faux-meat movement. These 'meats' from plant products differ from the meat substitutes produced for vegetarians by mimicking more closely the characteristics of real meat (taste, texture and sizzle on the grill), so the product will be attractive to meat eaters. All of these new approaches address the perceived failures of the high-input agriculture systems that produce today's food supply. They are promoted as being sustainable, environmentally friendly, encouraging biodiversity, reducing damage to the environment and producing healthier food.

Can these new approaches replace high-input conventional agriculture and feed the 9.7 billion people expected in 2050? Most of these approaches have positive attributes, but there are often negative ones as well and the negatives (usually completely overlooked by the promoters) may seriously limit their contribution to feeding the world. Many of these new agricultural systems emphasize activities oriented towards sustainability and minimizing environmental damage and end up sacrificing yield (Ponisio et al., 2015). Banning the use of synthetic fertilizers, especially N, puts a limit on maximum production. Erisman et al. (2008) estimated that synthetic N from the Haber–Bosch process was responsible for feeding 48% of the world population. Trying to feed 9.7 billion people with low-productivity agriculture creates problems and the solution to those problems often contravenes the stated goals of alternative agricultural systems. Low yields require more land under cultivation to achieve the same level of total production, but the damage to the environment associated with increasing cropland is well documented. Low yields would probably increase the cost of food, further exacerbating class differences in society. It would seem that the potential benefits of low-yield agricultural systems would not counterbalance the effects of the low yield in the battle to provide adequate food for future populations unless significant changes are made in food consumption patterns. Muller *et al.* (2017) suggested that alternative systems could feed the world if their lower yields were accompanied by reductions in food wastage and consumption of animal products, which would allow the land used to produce feed to shift to food production.

Many of these systems, especially mixed-cropping systems (polycultures), are labour intensive and the necessary labour supply may not be available, given the long-term trend for people to leave farming rather than embrace it. In some cases, complicated cropping systems would limit farm size and total income per farm, which would require higher prices for their production to be an economic success. Some of these proposed systems may actually mire the practitioners in continual poverty. Allen (2009) reported that small farms (2 ha (4.94 acres)) in the USA had larger net income per hectare than large farms (15,581 ha (38,485 acres)) (US$2902 ha^{-1} versus US$52 ha^{-1}, respectively). The total net income from the small farm (2 × US$2902) did not provide a living wage, while the large farm did (15,581 × US$52). This example, meant to illustrate the superiority of the small farm, actually illustrates perfectly how they are poverty traps. Many of these systems represent a return to traditional agriculture similar to that practised before the beginning of the high-input era, but 'nowhere has there been a need or serious desire, except amongst a privileged few, never full-time farmers, to return to the traditional farming practices left behind' (Fischer and Conner, 2018) – practices that led Jaclyn Moyer to lament 'what nobody told me about small farming: I can't make a living' (Moyer, 2016). When 91-year-old Ruth Parker was asked if she would go back to the old farm life, she answered: 'No. Frozen food and my microwave do just fine' (Kraig, 2017, p. 254). As these quotes suggest, attracting large numbers of participants to many of these forms of agriculture may be difficult, limiting their adoption as mainstream production systems.

The one new approach that has the potential, in my opinion, to contribute significantly to feeding the world in 2050 is the development of 'meat' made entirely from plant sources. The leading companies in the development of these products are Impossible Foods and Beyond Meat. Growth of meat from animal cells in culture is potentially another new source of protein, but this process only recently ventured out of the laboratory when lab-grown chicken nuggets were approved for sale in Singapore in late 2020 (O'Dowd and Hagan, 2020). The search for meat substitutes is not new. In the early 1900s, John Kellogg, of corn flakes fame, introduced the meat substitute 'protose' (a mixture of peanut butter and mashed beans) (Mansky, 2019). It is not surprising, given the composition, that protose did not catch on with consumers.

Today's faux meats are much closer to the real thing. The Impossible burger actually 'bleeds' as a result of the leghaemoglobin it contains and sizzles and browns when you throw it on the grill (Mansky, 2019). Impossible burgers are now available in fast-food outlets and Starbucks is selling a breakfast sandwich containing faux sausage. If they are accepted widely by the public, the decrease in animal numbers will reduce the land needed for feed production. Roughly 33.6 million ha (83 million acres) of maize were grown in the USA in recent years (2015–2019) (NASS, 2020) with roughly half of the production used for feed, while the 33.6 million ha (83 million acres) of soybean contributed high-protein feed supplements to the animal industry. Reduction in feed demand would release a significant proportion of this area for other uses. These unused hectares would provide badly needed flexibility – flexibility to respond to climate change by moving production to more hospitable environments, to shift to new food crops and to focus the remaining production on the most productive soils, thereby increasing yield and reducing environmental damage.

Economic success of many of these alternative systems with their low productivity and high labour requirements depends upon higher prices for their products. This is not a major issue as long as these systems are serving niche markets, i.e. there are consumers willing to pay a premium, for example, for vegetables from organic farms. However, higher prices extended to the entire food supply may not be acceptable if they create food insufficiency problems in lower-income segments of society.

Vertical farming operations are limited by high costs, primarily the high start-up costs and the high cost of electricity for lights to grow the crops. Consequently, they tend to focus on growing lettuce and other greens for high-end restaurants. One estimate suggested that the electricity cost for a loaf of bread from wheat grown indoors would be US$11 (Jurgens, 2020). Replacing the sun with electric lights, even if they are highly efficient light-emitting diodes (LEDs), does not seem to be the best approach to feeding a steadily growing population.

Widespread adoption of some of these approaches to food production may require large social changes that are a complete reversal of current long-standing trends including an increase in the number of farm workers, a shift from the convenience of processed food to the more labour-intense food preparation in the home kitchen and a lower standard of living for farmers. Societies in developed countries may find it difficult to make these adjustments, which would further limit the adoption of these systems.

Sustainability is often said to be the fundamental difference between these alternative systems and conventional high-input agricultural systems. The alternative systems are touted as being sustainable, while the high-input systems in use today are represented as unsustainable. What is sustainability?

One definition of sustain in my dictionary is 'to keep in existence, maintain or prolong'. A second definition is 'to endure, withstand'. These definitions suggest that sustainability is a very simple concept, but when applied to agricultural production systems it is anything but simple. Whether or not a cropping system is sustainable depends upon its effect on the environment (does it damage the environment), on its economic success (it will not survive if it does not show a profit) and on the availability of the resources needed to support the system. Modern high-input agriculture depends on adequate supplies of fertilizers, herbicides, pesticides, seeds, irrigation water (in some situations), machines and fuel to operate these machines and support global trade in agricultural commodities. An adequate supply of cheap energy underlies many of these inputs. The levels of inputs for many alternative systems are much lower, suggesting that they are, in fact, more sustainable.

Sustainability is difficult to evaluate in the short term. Which cropping system should be adopted to provide sustainability? This question can only be answered in its fullest form in retrospect; predicting future sustainability is difficult. The maize–soybean rotation has been common in the Midwestern maize belt in the USA for 70 years or more. This is a high-input system in all respects and the yields are still increasing. Is this system sustainable? It has been 'kept in existence', it has been maintained and its productivity has increased, so one could argue that it is sustainable. Others would argue that since it requires high levels of petroleum-based inputs (fuel, fertilizer, herbicides and pesticides) and since chemical fertilizers are (supposedly) destroying the soil, it is not sustainable. Excessive erosion also contributes to the unsustainability of the maize–soybean cropping system. Does the adoption of no-till planting techniques and cover crops that reduce erosion make it sustainable? The system still needs high inputs, so perhaps it is only more sustainable than the original. Varying degrees of sustainability further complicate any evaluation of the concept. Whether or not it is sustainable depends upon one's perspective.

Sustainability is, at its core, a poorly defined concept. As such, it provides little useful guidance, beyond its use as a feel-good buzzword, for defining the cropping systems of the future. In my opinion, a concept that cannot be clearly defined and measured is not a useful concept.

It seems unlikely, in my opinion, that these alternative systems will make a significant contribution to the food supply by 2050. The problems (negatives) discussed previously will likely prevent these systems from scaling up beyond being niche suppliers. The positives associated with these alternative systems are, at best, long-term benefits and, at worst, non-existent. Proponents of these systems often tout their sustainability – they are sustainable and conventional agriculture is not. Measuring/defining/predicting sustainability is, in the short term, difficult and often depends upon the perspective of the definer. Only time will tell which of these systems are truly sustainable. The one new form of agriculture that could have a major effect on our capacity to feed the world in 2050 is the development of faux meat. Eliminating a significant proportion of animal agriculture would decrease the land area used for feed production, which could have profound effects on food production systems.

The Search for the Silver Bullet – A Futile Quest?

Grain crop producers are constantly searching for higher yields. The question from producers that I encountered most frequently in my nearly 50 years of research at the University of Kentucky with soybean and other grain crops was 'what can I do to increase yield?'. Those producers were not interested in the tried-and-true fundamental management practices that provide the basis for high yield, they wanted something new. They were searching for a silver bullet: new technology that would dramatically increase their yield. A silver bullet that could be implemented by modifying their standard management programme, by applying the new technology as a seed treatment, applying it to the soil or by spraying it on the plant. The silver bullet they want must increase yield; simply improving the efficiency of production (same yield with fewer inputs), which improves the producer's bottom line, does not count as a silver bullet.

This search for something completely new implies that nearly 400 years of research on how plants grow has not completely determined the requirements for plant growth. Early scientists thought that plants got their sustenance from the soil. Jan van Helmont grew a substantial willow tree in a pot in 1648 and found no decrease in the weight of the soil in the pot, but he mistakenly concluded that plants got their sustenance from water. From this humble beginning, researchers eventually discovered photosynthesis and went on to investigate plant growth at ever-increasing levels of sophistication and detail, starting with whole plants and progressing to plant parts, plant processes, cells, organelles and, recently, down to the DNA that controls growth. This research described the plant processes responsible for growth (photosynthesis, respiration, N metabolism and protein synthesis, etc.), their nutrient requirements and the effect of environmental conditions on them. Crop physiologists investigated the growth of crop communities in the field; they studied the yield production process and how crops responded to management practices. It is hard to imagine that all of this research completely missed a 'silver bullet' that would significantly increase yield if applied to a crop growing in the field.

We can never say for sure that we know everything there is to know about how plants and crops grow, but if there is a silver bullet hiding somewhere in this system, it will take very detailed, careful, basic research to find it. We cannot say for sure that there is no silver bullet, but finding one, in my opinion, is a gamble with very long odds. Producers may be better off focusing on best management practices – practices solidly based on research – and applying them correctly to produce high yields.

Yield contests perfectly illustrate the search for the silver bullet in action. Originally organized by land-grant universities and commodity groups to encourage producers to use best management practices in their fields, they have come to symbolize the search for higher and higher yields. Producers marvel over the extremely high yields, asking the winners 'what did you do to get those yields?', but they do not want to hear about best management practices, they want the silver bullet. Producers of super-high yields are accorded the status usually reserved for rock stars as they travel the country describing how they produced such high yields.

Yield contests and super-high yields also illustrate the futility of the search for the silver bullet. I first encountered the record-yield syndrome when I went to the University of Illinois to start graduate school in 1965. A local farmer harvested 5375 kg soybean ha^{-1} (80 bu acre^{-1}) the previous autumn, a yield that was 3.2 times the average state yield (1680 kg ha^{-1} (25 bu acre^{-1})) (NASS, 2020), and the Agronomy Department was buzzing over what was responsible for that astounding yield. Some professors even used soil from the high-yielding field as fertilizer on their plots to no avail. The factors responsible for that high yield were never identified.

Contest and super-high yields increased steadily since then, with recent records reaching 12,766 kg ha^{-1} (190 bu acre^{-1}) for soybean (Anonymous, 2019) and 38,629 kg ha^{-1} (616 bu acre^{-1}) for maize (Spiegel, 2020). Each new record creates new theories to explain such high yields. Interestingly, many of the theories focused on the soil instead of the plant; no one wondered what caused the increased rate or duration of photosynthesis that had to occur to produce the high yield. On some occasions the high yields were simply attributed to a productive soil, a good variety, high fertility, plenty of water and control of weeds, insects and diseases – in short, all of the best management practices that are well-known components of high-yielding grain cropping systems. This answer was not satisfying because following those practices does not usually produce super-high yields. The causes of all these super-high yields were never determined to the point that they could be duplicated at other locations. If the cause of these super-high yields was reliably known, it would be quickly adopted by growers (assuming it was economically viable and could be practically applied to large acreages) and eventually become part of the standard management package for that crop and yields would increase dramatically. This has not happened.

Why some yields are super high is not known. I speculated in an article in 1982 (Egli, 1982) that record yields may occur when some rare combination of favourable environmental conditions comes together with best management practices. I still think it is a good theory, but it does not really tell us what it takes to produce super-high yields, i.e. it does not identify the silver bullet, and it does not explain why record yields sometimes occur in the same place for several years. Evaluation of yield contest results and super-high yields has not found a silver bullet.

Perhaps the bottom line is that there is no silver bullet that will dramatically increase yield. Perhaps the search for a silver bullet is a futile quest that distracts producers from aspects of crop management that actually influence yield. Making this statement does not mean that there never will be a silver bullet. Future research may well identify some unknown factor that is limiting crop growth that can be modified by management to dramatically increase yield. The new management practice will come from careful research, not a haphazard, trial-and-error search for a yield enhancer.

The silver bullet mentality has some negative aspects beyond distracting the producer, especially when producers try to emulate the yields of the stars. Some of the management practices associated with super-high yields do not represent good stewardship of the land and may damage the environment. For example, many contest winners and producers of super-high yields recommend multiple applications of fungicides and insecticides to 'keep the plant healthy' without considering the presence or absence of disease or insects. Indiscriminate use of these chemicals is not a good idea; it encourages development of resistance in the target organisms, which will eventually reduce the pesticide's effectiveness. Widespread indiscriminate use also attracts the attention of activists who want to ban pesticides because they damage the environment. Reckless overuse of pesticides may result in them not being available or effective when they are needed.

Record yield producers often recommend the use of N fertilizer on soybean, following the belief that the nodules on a soybean plant cannot fix enough N_2 to support high yields (Cafaro La Menza *et al.*, 2019). Soybean is a perfectly good legume that can produce high yields without N fertilizer. There is no need to convert it to a non-legume by applying copious amounts of N fertilizer when the damage that excess N does to the environment is well known. Discussions of abandoning the legume function of soybean come at the same time that others are attempting to establish N_2 fixation in cereals (Dent and Cocking, 2017) so they will require less N fertilizer, illustrating the illogical position of some soybean scientists.

The focus on extra-high yields encourages industry to develop products to stimulate plant growth and increase yield. The silver bullet mentality provides a ready market for a variety of miracle products that range from out and out magic to those that have some shred of scientific underpinning. These products often don't provide repeatable yield increases in replicated field trials and when they do, the increase is small. They are not the silver bullet, but they soak up producers' money and attention, and may distract them from doing a good job of applying the best management practices.

In my opinion, the search for a silver bullet is futile. It has not been found yet and the odds are that it will not be found. The search for a silver bullet is not only futile, it may reduce yield when producers neglect the best management practices when producing their crop. The requirements for high yield are reasonably well understood; producers are better off focusing on them and worrying more about their profit margin than record yields.

Summary

This chapter focused on the future. How will humankind deal with the challenges and opportunities they will face in the next 30 years? Our analysis (Fig. 5.1) suggests that the argument, put forward by some, that there is enough food available today to feed the population in 2050 if it was evenly distributed is not true. Production of grain crops will have to increase in the next 30 years, but the increase may not have to be as large as many of the current estimates suggest. We discussed new techniques and new technologies in this chapter that may contribute to meeting the world's food needs in 2050.

Du Bois and Mintz (2008) defined several perspectives on the issues associated with maintaining a well-fed world through 2050, providing a useful framework to evaluate the various options.

1. Steadfast optimists: 'Confident that human ingenuity will meet the challenge through rapid technological growth and cultural adaptation. Their view is that in 2050 humans will be able to produce enough food ... to satisfy everyone's basic needs' (Du Bois and Mintz, 2008, p. 303). A closely related group are the cautious optimists who 'argue that the rate of population growth must continue to decline, that we must give farmers adequate support, that we must devise technologies to sustainably improve yields or expand cropland, and that we must boost the incomes of the world's poorest' (Du Bois and Mintz, 2008, p. 304).

2. Social justice: Feeding the 9.7 billion people in 2050 requires far-reaching social solutions rather than technical ones. 'Population growth is of little concern since malnutrition is seen as a consequence of social failings in the midst of planetary abundance. If poor farmers were not so exploited by large multinational corporations, governments and local elites, they could provide enough food for them and their offspring' (Du Bois and Mintz, 2008, p. 304). This approach favours a low-tech, labour-intensive form of farming that relies on local wisdom of time-honoured, largely pre-industrial farming practices producing a greater diversity of crops for local consumption.

3. Resource pessimists: They 'agree with the Social Justice advocates that some form of organic or ecologically integrated agriculture is needed, but they are very concerned about the growth in population and its effects on local natural and social environments. They see the food–population problem leading to wars over precious land, water, energy, and food with nature controlling population through poverty, disease, and starvation' (Du Bois and Mintz, 2008, p. 305). Their solution is to immediately stop population growth and maintain a population of 7 billion.

4. Neo-Malthusian biotechnologists: 'Capitalistic biotechnology, if left unencumbered, will succeed in feeding the world. This view is heavily promoted by companies invested in genetic engineering' (Du Bois and Mintz, 2008, p. 306).

All of these options are part of the global debate over how to feed the world in 2050, but they present stark differences in how to accomplish that goal. The options range from complete dependence on biotechnology as practised by large corporations to a world populated by peasant farmers doing agriculture the old way. Our discussions in this chapter suggest that the social justice approach will not work. A world of small farmers feeding their families does not fit the reality of a world where half the population lives in cities. The movement of people away from the farm that accelerated during the Industrial Revolution and continues unabated to this day suggests that, by and large, people are not interested in being small farmers condemned to a life of drudgery and poverty.

The goal of resource pessimists to limit population growth is, in a sense, being met as population growth rates in many countries have already dropped below replacement rates. These slowing growth rates will not, however, limit world population to 7 billion but it may be less in 2050 than the United Nations' (2019) median estimate of 9.7 billion people.

In my opinion, the steadfast and cautious optimists, in cooperation with the biotechnologists, provide the clearest path towards a well-fed world in 2050. The median estimate from the United Nations (2019) projects a population increase of 25% (14% for the low estimate) by 2050, which is substantially less than the 47% increase in the previous 30 years (1990–2020) when production increased fast enough to create a worldwide obesity pandemic. The challenge we face in 2020 seems to be substantially less than the one we conquered in the previous 30 years.

New technology, including biotechnology and new techniques to manipulate the plant genome, precision agriculture and big data, is likely to make significant contributions to higher yields and/or greater production efficiencies in the future. The development of faux meats could displace animal agriculture and its inefficiencies and increase the land available for food production. Surely, the combination of the traditional tried-and-true approaches to increasing agricultural production, which have served us well since the beginning of the high-input agriculture era in the 1930s, and the new technologies that are coming online, technology that is in some cases truly amazing, provides many opportunities to increase food production. Environmental damage resulting from this technology will have to be eliminated to maintain high-yielding agricultural systems. The systems that feed us in 2050 may be unrecognizable by today's standards, but it is likely we will be well fed and happy, assuming we can overcome the challenge of climate change.

Climate change represents a major stumbling block on the path to a well-fed world in 2050. How large the stumbling block will be is, in my opinion, not yet clear. I do not believe that we know all the ramifications of the increase in temperature on climate. Meeting the challenges of a rising temperature will be difficult, but if higher temperatures are accompanied by significant droughts in key agricultural regions, as many predict, the problem will be much more difficult.

I remain an optimist because, as the great American author William Faulkner put it: 'I believe that man will not merely endure: he will prevail. He is immortal, not because he alone among creatures has an inexhaustible voice, but because he has a soul, a spirit capable of compassion and sacrifice and endurance' (Faulkner, 1950).

General Summary

The thesis of this book is that a greater knowledge of the fundamental processes controlling the growth of crop communities and the production of yield (i.e. crop physiology) will help us be better managers; managers who produce maximum yield efficiently with minimal damage to the environment. Crop physiology provides the basis for understanding the management practices used in grain production and it is this understanding that makes us better. To this end, we first discussed basic plant growth processes, followed by an evaluation of the growth of crop communities, and finally we used that knowledge to understand the 'why' component of important management practices.

Photosynthesis of the crop community is the process that produces yield; almost all of the plant tissues – leaves, stems, roots and seeds – that make up the mature crop come from photosynthesis. When crop managers select, for example, planting date, plant population and row spacing they are actually managing photosynthesis. They are trying to create an environment that will maximize photosynthesis. Almost any management practice or environmental effect on yield is expressed through its effect on photosynthesis. Fertilizing a crop with N often increases yield, but the N per se does not increase yield, it simply simulates photosynthesis. The same can be said for irrigation; the added water increases yield by increasing photosynthesis. The yield–photosynthesis association is not perfect; yield can decrease without a change in photosynthesis. For example, when high temperature reduces pollination or when insects feed on reproductive structures, the yield reduction is not directly related to a lack of photosynthesis. The exceptions do not disprove the rule: yield at physiological maturity is, by and large, determined by photosynthesis, either the rate or the duration or both.

The physiological processes and the basic relationships that support basic management decisions are reasonably well understood, which makes it possible to generally predict the outcome of many management decisions. Understanding why crops respond to management is an invaluable aid to being a better manager. In spite of this knowledge, we continue to investigate management practices experimentally, as if for some unknown reason we will get a different result. We should keep in mind that one definition of insanity is repeating the same action and expecting a different result. Agronomists generally have great faith in data; they are very hesitant to extrapolate. I think this attitude is, at least partially, a result of the ease of doing field research. Running experiments to compare several row spacings is not complicated or expensive, so we do it over and over again. The fact that we usually get the same response makes no impression on us. I think the need to obtain funding for their research from commodity organizations that favour applied research also encourages agronomists to constantly run experiments that reinvent the wheel. This repetitious research may, in some cases, also be an indication of a lack of imagination among the practitioners.

Comparison of biologists and agronomists, in this respect, is, I think, enlightening. A biologist will study a few residents of the stream in his back yard and tell you how the whole world works. An agronomist, when asked a management question by a farmer from an adjacent county, will respond 'we found this response here, but we have not investigated this practice in your county'. I have, of course, exaggerated the extremes, but I believe the basic premise is true: agronomists are very reluctant to extrapolate or generalize, always worried about variety and environment interactions. Understanding the basic principles governing the response provides a stronger basis for generalizing. I hope that this book will encourage its readers to generalize more.

We cannot discuss crop management without thinking about the future. What does the future hold for crop physiologists and agronomists? The literature generally paints a dark and dismal picture, projecting large increases in yield needed to ensure a well-fed world in 2050. The truth seems to be a little less daunting (see Hunter *et al.*, 2017 for an interesting analysis of the 'crop production must double by 2050' projection). Dark and dismal is a good way to justify research funding, but it may not be entirely true. Our analysis of wheat and rice production through 2050 suggests a more manageable situation. A yield growth rate of one-half of the rate for the last 10 years was all that was needed to maintain per capita production of wheat and rice when the harvested area was held constant and the median estimate of population growth was used. The status quo was maintained with a significant reduction in the yield growth rate, a much more doable situation than having to increase the rate of yield growth. Perhaps feeding the world will not be as difficult as we think. If population starts declining by 2050 instead of increasing, as Bricker and Ibbitson (2019) suggest, feeding the world will get a lot easier.

The big challenge facing the world today and into the future is climate change. How will local climates change in response to global warming? How big will the impact be on agricultural production? How much warming will occur? Can we maintain productivity by changing cultural practices or will we need new varieties that are much more stress tolerant? These questions will be answered as climate change progresses. Humanity still seems reluctant to limit the emission of greenhouse gases, suggesting that the eventual disruption of agricultural productivity may be severe. A basic knowledge of crop growth and the principles regulating the response to management gleaned from this book will hopefully help with the adjustment.

The new technologies that are starting to come online will help manage the transition to a hotter world. The application of molecular biology to variety development, precision agriculture, big data and artificial intelligence will contribute to maintaining agricultural productivity in the future. The challenge is to determine how to use these exciting new technologies and, again, the relationships discussed in this book will be helpful.

Some of the technology on the horizon that will potentially play a huge role in feeding the world in 2050 is not related directly to grain crop production. The development of faux meats from plant sources and meat grown from animal cells in culture could have far-reaching effects on food supplies and agricultural production systems. The goal of Patrick Brown, CEO of Impossible Foods, the company that markets the Impossible Whopper at Burger King, is to completely replace all animals in the food system by 2035 (Little, 2019, p. 189). Changes of this magnitude will completely disrupt agricultural systems and make the challenge of feeding the world much simpler. It is hard to imagine the agriculture scene in the USA without the 67 million ha (166 million acres) of maize and soybean grown largely for animal feed. Large changes in crops grown have occurred in the past. For example, the shift from animal to tractor power greatly reduced hay and oat crops, world soybean production grew from zero in 1900 to 74 billion ha (183 billion acres) in 2000 (FAOSTAT, 2020), oilseed rape did not exist in 1979 and now it occupies 37.6 million ha (92.9 million acres) globally. These changes suggest that society could make large changes in the crops grown in the future. Barring large increases in prosperity in underdeveloped countries with food shortages that would allow them to buy food on the world market, the shift may involve reductions in the area under cultivation in some countries, similar to those occurring in Japan.

We cannot predict with any certainty what will happen in the future, but it seems to me that the odds of beating the Malthusian trap once again are good. Good only if we can successfully manage climate change.

Appendix

Table A1. Conversion table.

Distance	Temperature		
1 cm (centimetre) = 0.394 in = 10 mm (millimetre)	°C (degree Celsius) = 5/9 × (°F − 32)		
1 in (inch) = 2.54 cm = 25.4 mm	°F (degree Fahrenheit) = (9/5 × °C) + 32		
1 m (metre) = 3.281 ft	38.6°C = 101.5°F		
1 ft (foot) = 0.3048 m = 12 in	38.9°C = 102.0°F		
	39.4°C = 103.0°F		
1 km (kilometre) = 3281 ft = 0.6214 mile	40.0°C = 104.0°F		
1 mile = 1609.344 m = 1.609 km = 5280 ft	40.6°C = 105.0°F		

Area	Mass		
1 m² (square metre) = 10.7639 ft²	1 g (gram) = 0.03527 oz		
1 ft² (square foot) = 0.0929 m²	1 oz (ounce) = 28.35 g		
1 cm² (square centimetre) = 0.155 in²	1 kg (kilogram) = 2.205 lb = 35.28 oz		
1 in² (square inch) = 6.4516 cm²	1 lb (pound) = 0.4535 kg = 453.5 g = 16 oz		
1 ha (hectare) = 2.471 acre = 10,000 m² = 107,639 ft²	50 kg = 110 lb		650 kg = 1433 lb
1 acre = 0.4047 ha = 43,560 ft² = 4047 m²	100 lb = 45.4 kg		900 lb = 408 kg
	400 kg = 882 lb		1000 lb = 454 kg
1 km² (square kilometre) = 0.3861 square mile = 100 ha	450 kg = 992 lb		1100 lb = 499 kg
1 sq mile = 2.589 km² = 640 acre	500 kg = 1102 lb		1200 lb = 544 kg
	550 kg = 1213 lb		1300 lb = 590 kg
	600 kg = 1323 lb		1400 lb = 635 kg

Continued

© D.B. Egli 2021. *Applied Crop Physiology: Understanding the Fundamentals of Grain Crop Management* (D.B. Egli)
DOI: 10.1079/9781789245950.000a

Table A1. Continued.

Volume

1 US fl oz (US fluid ounce) = 0.0296 l
1 l (litre) = 0.264 US gal
1 US gal (US gallon) = 3.785 l = 128 US fl oz
1 bu (bushel) = 36.4 l
1 ft³ (cubic foot) = 0.0283 m³
1 m³ (cubic metre) = 14.046 ft³

Table A2. Scientific and common names of all plant species mentioned in the text.

Common name	Scientific name	Common name	Scientific name
Barley	*Hordeum vulgare* L.	Oilseed rape	*Brassica napus* L., *Brassica campestris* L.
Bean	*Phaseolus vulgaris* L.	Pea	*Pisum sativum* L.
Broad bean[a]	Vicia faba L.	Pearl millet	*Pennisetum glaucum*
Buckwheat	*Fagopyrum esculentum*	Pigeon pea	*Cajanus cajan* L. Millsp.
Canola	*Brassica napus* L., *Brassica campestris* L.	Potato	*Solanum tuberosum* L.
Cassava	*Manihot esculenta* Crantz	Potato bean	*Apios americana*
Chia	*Salvia hispanica*	Quinoa	*Chenopodium quinoa*
Chickpea	*Cicer arietinum* L.	Rape	*Brassica napus* L., *Brassica campestris* L.
Cotton	*Gossypium hirsutum* L.	Rice	*Oryza sativa* L.
Cowpea	*Vigna unguiculata* (L.) Walp	Rocket (arugula)	*Eruca vesicaria* ssp. *sativa*
Field pea	*Pisum arvense* L.	Rye	*Secale cereale* L.
Flax	*Linum usitatissimum* L.	Sesame	*Sesamum indicum* L.
Grain amaranth	*Amaranthus* spp.	Sorghum	*Sorghum bicolor* (L.) Moench
Groundnut (peanut)	*Arachis hypogaea* L.	Soybean	*Glycine max* (L.) Merrill
Hemp	*Cannabis sativa*	Spinach	*Spinacia oleracea* L.
Intermediate wheat grass	*Thinopyrum intermedium*	Sugarbeet	*Beta vulgaris* (L.) ssp. *vulgaris*
Kale	*Brassica oleracea* L.	Sugarcane	*Saccharum officinarum* L.
Lentil	*Lens culinaris* Medikus	Sunflower	*Helianthus annuus* L.
Lettuce	*Lactuca sativa*	Tall fescue	*Festuca arundinacea*

Continued

Table A2. Continued.

Common name	Scientific name	Common name	Scientific name
Lucerne (alfalfa)	*Medicago sativa* L	Teff	*Eragrostis tef*
Lupin	*Lupinus mutabilis*	Tomato	*Solanum lycopersicum*
Maize (corn)	*Zea mays* L.	Triticale	× *Triticosecale Wittmack*
Millet	*Panicum miliaceum* L.	Vernonia	*Vernonia galamensis*
Mung bean[b]	*Vigna radiata*	Wheat	*Triticum* spp.
Mustard	*Brassica juncea* (L.) Czern et Coss	White lupin	*Lupinus albus*
Oat	*Avena sativa* L.		

[a]Faba bean.
[b]Black or green gram.

References

Abendroth, L.J., Elmore, R.W., Boyer, M.J. and Marlay, S.K. (2011) *Corn Growth and Development*. PMR 1009. Iowa State University Extension, Ames, Iowa.

Abendroth, L.J., Woli, K.P., Meyers, A.J.W. and Elmore, R.W. (2017) Yield-based corn planting date recommendation windows for Iowa. *Crop, Forage & Turfgrass Management* 3, 1–7.

Allen, M.A. (2009) Agroecology, small farms and food sovereignty. *Monthly Review* 61, 102–113.

Anonymous (2017) Dupont–Pioneer announces 69–CRM corn hybrid. *MidAmerica Farmer Grower* (37), 7. MidAmerica Farm Publications, Inc., Perryville, Missouri.

Anonymous (2019) Randy Dowd breaks soybean yield world record. *AGDAILY*, Crops News, 23 September 2019. Available at: https://www.agdaily.com/crops/randy-dowdy-soybean-yield-record (accessed 15 March 2021).

AOSA (2009) *Seed Vigor Handbook*. Contribution No. 32. Association of Official Seed Analysts, Ithaca, New York.

AOSA (2019) *Rules for Testing Seeds*. Vol. 2, *Principles and Procedures*. Association of Official Seed Analysts, Stillwater, Oklahoma.

APPA (2016) *2015–2016 National Pet Owners Survey*. American Pet Products Association, Greenwich, Connecticut.

Austin, R.B., Bingham, J., Blackwell, R.D., Evans, L.T., Ford, M.A., Morgan, C.L. and Taylor, M. (1982) Genetic improvement in winter wheat yields since 1900 and associated physiological changes. *Journal of Agricultural Science* 94, 685–689.

Bassetti, P. and Westgate, M.E. (1993) Water deficit affects receptivity of maize silks. *Crop Science* 33, 279–282.

Belamkar, V., Wenger, A., Kalberer, S.R., Bhattacharya, V.G., Blackmon, W.J. and Cannon, S.B. (2015) Evaluation of phenotypic variation in a collection of *Apios americana*: an edible tuberous legume. *Crop Science* 55, 712–726.

Bewley, J.D., Bradford, K.J., Hilhorst, H.W.M. and Nonogaki, H. (2013) *Seeds: Physiology of Development, Germination and Dormancy*, 3rd edn. Springer, New York.

Bjerga, A. (2012) Canada's new corn belt attracts hot money to bargain farmland. Available at: https://www.farmlandgrab.org/post/view/21281-canadas-new-corn-belt-attracts-hot-money-to-bargain-farmland (accessed 4 May 2021).

Black, M., Bewley, J.D. and Halmer, P. (2006) *The Encyclopedia of Seeds: Science, Technology and Uses*. CAB International, Wallingford, UK.

Boote, K.J. (1982) Growth stages of peanut (*Arachis hypogaea* L.). *Peanut Science* 9, 35–40.

Boyer, C.N., Stefanini, M., Larson, J.A., Smith, S.A., Mengistu, A. and Bellaloui, N. (2015) Profitability and risk analysis of soybean planting date by maturity group. *Agronomy Journal* 107, 2253–2262.

Brevedan, R.E. and Egli, D.B. (2003) Short periods of water stress during seed filling, leaf senescence and yield of soybean. *Crop Science* 43, 2083–2088.

Bricker, D. and Ibbitson, J. (2019) *Empty Planet: The Shock of Global Population Decline*. Crown, New York.

Bunting, A.H. (1971) Productivity and profit or is your vegetative phase really necessary? *Annals of Applied Biology* 67, 265–285.

Bunting, A.H. (1991) Feeding the world in the future. In: Spedding, C.R.W. (ed.) *Frean's Principles of Food and Agriculture*, Vol. 17. Blackwell Scientific Publications, Oxford.

Cafaro La Menza, N., Monzon, J.P., Specht, J.E, Lindquist, J., Arkebaver, T.J., Graef, G. and Grassini, P. (2019) Nitrogen limitation in high yield soybean: seed yield, N accumulation, and N use efficiency. *Field Crops Research* 237, 74–81.

Caruso, R. (1997) *The Brazilian Cerrado: Development, Conservation and Sustainability*. Fundacao Cargill, Campinas, Brazil.

Cassman, K.G. and Liska, A.J. (2007) Food and fuel for all: realistic or foolish? *Biofpr: Biofuels, Bioproducts & Biorefining* 1, 18–23.

Charles-Edwards, D.A. (1982) *Physiological Determinants of Crop Growth*. Academic Press Australia, North Ryde, Australia.

Cheng, A. (2018) Review: Shaping a sustainable food future by re-discovering long-forgotten ancient grains. *Plant Science* 269, 136–142.

Clark, M.A., Domango, N.G.G., Colgan, K., Thakrar, S.K., Tilman, D., *et al.* (2020) Global food system emissions could preclude achieving the 1.5 and 2°C climate change targets. *Science* 370, 705–708.

Cohen, J.E. (1995) *How Many People Can the Earth Support?* W.W. Norton, New York.

Collins, W.K., Russell, W.A. and Eberhart, S.A. (1965) Performance of two-ear type of cornbelt maize. *Crop Science* 5, 113–116.

Conford, P. and Dimbleby, J. (2001) *Origins of the Organic Movement*. Floris Books, Edinburgh, UK.

Conner, D.J., Loomis, R.S. and Cassman, K.G. (2011) *Crop Ecology, Productivity and Management*. Cambridge University Press, Cambridge.

Cooper, M., Gao, C., Leafgren, H., Tang, T. and Messina, C. (2014) Breeding drought tolerant maize hybrids for the US corn belt: discovery to product. *Journal of Experimental Botany* 65, 6191–6204.

Crooks, W. (1898) Address of the president before the British Association for the Advancement of Science, 1898. *Science* 8, 561–575.

CRA (2016) NASA Prediction of Worldwide Energy Resources (POWER). Climatology Resources for Agroclimatology. Available at: https://power.larc.nasa.gov (accessed 23 February 2021).

Daynard, T.B. and Duncan, W.G. (1969) The black layer and grain maturity in corn. *Crop Science* 9, 473–476.

Deffeyes, K.S. (2001) *Hubbert's Peak: The Impending World Oil Shortage*. Princeton University Press, Princeton, New Jersey.

Dent, D. and Cocking, E. (2017) Establishing symbiotic nitrogen fixation in cereals and other non-legume crops: the greener nitrogen revolution. *Agriculture and Food Security* 6, 7.

Despommier, D. (2010) *The Vertical Farm: Feeding the World in the 21st Century.* Thomas Dunne Books/St. Martin's Press, New York.

Diamond, J. (1987) The worst mistake in the history of the human race. *Discover Magazine* (May), 64–66.

Donald, C.M. (1962) In search of yield. *Journal of the Australian Institute of Agricultural Science* 28, 171–178.

Donald, C.M. (1968) The breeding of crop ideotypes. *Euphytica* 17, 385–403.

Du Bois, C.W. and Mintz, S.W. (2008) Soy's dominance and destiny. In: Du Bois, C.W., Tan, C.-B. and Mintz, S.W. (eds) *The World of Soy.* University of Illinois Press, Urbana and Chicago, Illinois, pp. 299–313.

Duncan, W.G. (1971) Leaf angles, leaf area, and canopy photosynthesis. *Crop Science* 11, 482–485.

Duncan, W.G., Shaver, D.L. and Williams, W.A. (1973) Insolation and temperature effects on maize growth and yield. *Crop Science* 13, 187–189.

Dunwell, J.M. (2010) Crop biotechnology: prospects and opportunities. *Journal of Agricultural Science* 149, 17–27.

Duvick, D.N. (2005) The contribution of breeding to yield advances in maize (*Zea mays*). *Advances in Agronomy* 86, 84–145.

Easterbrook, G. (2018) *It's Better than It Looks: Reasons for Optimism in an Age of Fear.* Hachette Book Group, New York.

Egan, T. (2006) *The Worst Hard Time: The Untold Story of Those Who Survived the Great American Dust Bowl.* Houghton Mifflin Co., New York.

Egli, D.B. (1982) Why don't 50-bushel beans make 70? *Soybean News* 37(3), 1.

Egli, D.B. (1991) W.G. Duncan – father of crop models. *Journal of Agronomic Education* 20, 165–167.

Egli, D.B. (1993a) Cultivar maturity and potential yield of soybean. *Field Crops Research* 32, 147–158.

Egli, D.B. (1993b) Relationship of uniformity of soybean seedling emergence to yield. *Journal of Seed Technology* 17, 22–28.

Egli, D.B. (2004) Seed-fill duration and yield of grain crops. *Advances in Agronomy* 83, 243–279.

Egli, D.B. (2008a) Comparison of corn and soybean yield trends in the United States: historical trends and future prospects. *Agronomy Journal* 100, S79–S88.

Egli, D.B. (2008b) Soybean yield trends from 1972 to 2003 in mid-western USA. *Field Crops Research* 106, 53–59.

Egli, D.B. (2010) Soybean reproductive sink size and short-term reductions in photo-synthesis during flowering and pod set. *Crop Science* 50, 1971–1977.

Egli, D.B. (2011) Time and the productivity of agronomic crops and cropping systems. *Agronomy Journal* 103, 743–750.

Egli, D.B. (2015a) Is there a role for sink size in understanding maize population–yield relationships? *Crop Science* 55, 1–10.

Egli, D.B. (2015b) Pod set in soybean: investigations with SOYPODP, a whole plant model. *Agronomy Journal* 107, 1–12.

Egli, D.B. (2017) *Seed Biology and the Yield of Grain Crops*, 2nd edn. CAB International, Wallingford, UK.

Egli, D.B. (2019) KNMAIZE – a model to investigate the dynamics of kernel set in maize. *Agronomy Journal* 111, 1720–1727.

Egli, D.B. and Bruening, W.P. (2000) Potential of early-maturing soybean cultivars in late plantings. *Agronomy Journal* 92, 532–537.

Egli, D.B. and Bruening, W.P. (2006) Temporal patterns of pod production and pod set in soybean. *European Journal of Agronomy* 24, 11–18.

Egli, D.B. and Cornelius, P.L. (2009) A regional analysis of the response of soybean yield to planting date. *Agronomy Journal* 101, 330–335.

Egli, D.B. and Hatfield, J.L. (2014a) Yield gaps and yield relationships in central US soybean production systems. *Agronomy Journal* 106, 560–566.

Egli, D.B. and Hatfield, J.L. (2014b) Yield and yield gaps in central US corn production systems. *Agronomy Journal* 106, 2248–2254.

Egli, D.B. and Rucker, M. (2012) Seed vigor and the uniformity of emergence of corn seedlings. *Crop Science* 52, 2774–2782.

Egli, D.B. and Yu, Z.-W. (1991) Crop growth rate and seeds per unit area in soybean. *Crop Science* 31, 434–442.

Egli, D.B., Pendleton, J.W. and Peters, D.B. (1970) Photosynthesis rate of three soybean communities as related to carbon dioxide levels and solar radiation. *Agronomy Journal* 62, 411–414.

Egli, D.B., Wiralaga, R.A. and Ramseur, E.L. (1987) Variation in seed size in soybean. *Agronomy Journal* 79, 463–467.

Egli, D.B., Hamman, B. and Rucker, M. (2010) Seed vigor and uniformity of seedling emergence in soybean. *Seed Technology* 32, 87–95.

Ehrlich, P.R. (1968) *The Population Bomb*. Ballantine Books, New York.

Engledow, F.L. and Wadham, S.M. (1923) Investigations of yield in cereals I. *Journal of Agricultural Science* 13, 390–439.

Erisman, J.W., Sutton, M.A., Galloway, J., Kliment, Z. and Winiwarter, W. (2008) How a century of ammonia synthesis changed the world. *Nature Geoscience* 1, 636–639.

Ermakova, M., Danila, F.R., Furbank, R.T. and von Caemmerer, S. (2020) On the road to C_4 rice: advances and perspectives. *The Plant Journal* 101, 949–950.

Evans, L.T. (1975) Crops and the world food supply, crop evolution and the origins of crop physiology. In: Evans, L.T. (ed.) *Crop Physiology: Some Case Histories*. Cambridge University Press, Cambridge, pp. 1–22.

Evans, L.T. (1993) *Crop Evolution, Adaptation and Yield*. Cambridge University Press, Cambridge.

Fabrizius, E.E. (1993) Predicting changes in soybean seed germination during storage under warehouse conditions. MS thesis, University of Kentucky, Lexington, Kentucky.

FAOSTAT (2020) Crops. Available at: http://www.fao.org/faostat/en/#data/QC (accessed 24 February 2021).

Faulkner, W. (1950) William Faulkner – banquet speech, 10 December. Available at: https://www.nobelprize.org/prizes/literature/1949/faulkner/speech/ (accessed 24 February 2021).

Fehr, W.R. and Caviness, C.E. (1977) *Stages of Soybean Development*. Special Report No. 80. Iowa State University, Ames, Iowa.

Fischer, T. and Conner, D.J. (2018) Issues for cropping and agricultural science in the next 20 years. *Field Crops Research* 222, 121–142.

Fischer, T., Byerlee, D. and Edmeades, G. (2014) *Crop Yields and Global Food Security: Will Yield Continue to Feed the World?* ACIAR Monograph No. 158. Australian Centre for International Agricultural Research, Canberra.

Foley, J.A., Ramankutty, N., Brauman, K.A., Cassidy, E.S., Gerber, J.S., *et al.* (2011) Solutions for a cultivated planet. *Nature* 478, 337–342.

Fraser, J., Egli, D.B. and Leggett, J.E. (1982) Pod and seed development on soybean cultivars with differences in seed size. *Agronomy Journal* 74, 81–85.

Gao, J., Shi., S., Dong, S., Liu, P., Zhao, B. and Zhang, J. (2017) Grain yield and root characteristics of summer maize (*Zea mays* L.) under shade stress conditions. *Journal of Agronomy and Crop Science* 203, 562–573.

Gardner, B.L. (2002) *American Agriculture in the Twentieth Century: How It Flourished and What It Cost.* Harvard University Press, Cambridge, Massachusetts.

Gardner, F.P., Pearce, R.B. and Mitchell, R.L. (1985) *Physiology of Crop Plants.* Iowa State University Press, Ames, Iowa.

GBD 2015 Obesity Collaborators (2017) Health effects of overweight and obesity in 195 countries over 25 years. *New England Journal of Medicine* 377, 13–27.

Gelinas, B. and Seguin, P. (2008) Evaluation of management practices for grain amaranth production in eastern Canada. *Agronomy Journal* 100, 344–351.

Glover, J.D., Reganold, J.P., Bell, L.W., Boreuitz, J., Brummer, E.C., *et al.* (2010) Increased food and ecosystem security via perennial grains. *Science* 328, 1638–1639.

Griliches, Z. (1957) Hybrid corn: an exploration in economics of technological change. *Econometrics* 25, 501–502.

Hager, T. (2008) *The Alchemy of Air: A Jewish Genius, a Doomed Tycoon, and the Scientific Discovery that Fed the World, but Fueled the Rise of Hitler.* Harmony Books, New York.

Ham, J.M. (2005) Useful equations and tables in micrometeorology. In: Viney, M.K. (ed.) *Micrometeorology in Agricultural Systems.* American Society of Agronomy–Crop Science Society of America–Soil Science Society of America, Madison, Wisconsin, pp. 533–560.

Hanway, J.J. (1963) Growth stages of corn (*Zea mays* L.). *Agronomy Journal* 55, 487–492.

Harlan, J.R. (1992) *Crops and Man*, 2nd edn. Crop Science Society of America, Madison, Wisconsin.

Hartwig, E.E. (1973) Variety development. In: Caldwell, B.E. (ed.) *Soybeans: Improvement, Production and Uses.* American Society of Agronomy, Madison, Wisconsin, pp. 187–210.

Hatfield, J.L., Boote, K.J., Kimball, B.A., Ziska, L.H., Izaurralde, R.C., *et al.* (2011) Climate impacts on agriculture: implications for crop production. *Agronomy Journal* 103, 351–370.

Hay, R.K.M. and Porter, J.R. (2006) *The Physiology of Crop Yield*, 2nd edn. Blackwell Publishing Ltd, Oxford.

Heatherly, L.G. (1999) Early soybean production system (ESPS). In: Heatherly, L.G. and Hodges, H.F. (eds) *Soybean Production in the Midsouth.* CRC Press, Boca Raton, Florida, pp. 103–118.

Heiser, C.B.J. (1973) *Seed to Civilization – The Story of Man's Food.* W.H. Freeman & Co., San Francisco, California.

Herbek, J.H. and Bitzer, M.J. (1988) *Soybean Production in Kentucky. Part III. Planting Practices and Double Cropping.* University of Kentucky, Lexington, Kentucky.

Hessor, L. (2006) *The Man Who Fed the World. Nobel Peace Laureate Norman Borlaug and His Battle to End World Hunger.* Durban House Publishing Company Inc., Dallas, Texas.

Hulse, J.H., Laing, E.M. and Pearson, O.E. (1980) *Sorghum and the Millets: Their Composition and Nutritive Value.* Academic Press, London.

Hunter, J.L., TeKrony, D.M., Miles, D.F. and Egli, D.B. (1991) Corn seed maturity indicators and their relationship to uptake of carbon-14 assimilate. *Crop Science* 31, 1309–1313.

Hunter, M.C., Smith, R.G., Schipanski, M.E., Atwood, L.D. and Mortensen, D.A. (2017) Agriculture in 2050: recalibrating targets for sustainable intensification. *Bioscience* 67, 386–391.

Jackson, W. (1985) *New Roots for Agriculture.* University of Nebraska Press, Lincoln, Nebraska.

Jadhav, R. (2017) Growing supply glut threatens worse to come for restive Indian farmers. *Reuters Economic News.* Available at: https://reuters.com/article/idUSL3N1J926J (accessed 24 February 2021).

Jamboonsri, W., Phillips, T., Geneve, R., Cahill, J. and Hildebrand, D. (2012) Extending the range of an ancient crop, *Salvia hispanica* – a new ω3 source. *Genetic Resources and Crop Evolution* 59, 171–178.

Jurgens, J. (2020) Vertical farming: how plant factories stack up against field agriculture. *AEM Newsletter.* Available at: https://www.aem.org/news/vertical-farming-how-plant-factories-stack-up-against-field-agriculture (accessed 24 February 2021).

Kantar, M., Tyl, C.T., Dorn, K.M., Zhang, X., Jungers, J.M., *et al.* (2016) Perennial grain and oilseed crops. *Annual Review of Plant Biology* 67, 703–729.

King, J. (1997) *Reaching for the Sun: How Plants Work.* Cambridge University Press, Cambridge.

Klingaman, W.K. and Klingaman, N.P. (2013) *The Year Without Summer: 1876 and the Volcano that Darkened the World and Changed History.* St. Martin's Press, New York.

Knott, C., Herbek, J.H. and James, J. (2019) Early planting dates maximize soybean yield in Kentucky. *Crop, Forage & Turfgrass Management* 5, 180185.

Kraig, B. (2017) *A Rich and Fertile Land: A History of Food in America.* Reaktion Books Ltd, London.

Kucharik, C.J. (2006) A multidecadal trend of earlier corn planting in the central USA. *Agronomy Journal* 98, 1544–1550.

Langer, R.H.M. and Hill, G.D. (1991) *Agricultural Plants,* 2nd edn. Cambridge University Press, Cambridge.

Large, E.C. (1954) Growth stages in cereals. *Plant Pathology* 3, 128–129.

Lee, C.D., Egli, D.B. and TeKrony, D.M. (2008) Soybean response to early and late planting dates in the mid-south. *Agronomy Journal* 100, 971–976.

Linderholm, H.W. (2006) Growing season changes in the last century. *Agriculture and Forest Meteorology* 137, 1–14.

Little, A. (2019) *The Fate of Food: What We'll Eat in a Bigger, Hotter, Smarter World.* Harmony Books, New York.

Malthus, T.R. (1993) *An Essay of the Principle of Population* (originally published 1798). Oxford University Press, Oxford.

Mansky, J. (2019) We're entering a new age of meatless meat today. But we've been here before. *Smithsonian Magazine.* Available at:

https://www.smithsonianmag.com/arts-culture/turn-century--meatless-meat-180972042 (accessed 24 February 2021).

Mooers, C.A. (1908) *The Soy Bean: A Comparison with Cowpea.* Bulletin No. 82. Tennessee Agricultural Experiment Station, Knoxville, Tennessee.

Moyer, J. (2016) What nobody told me about small farming: I can't make a living. Available at: https://www.salon.com/2015/02/10/what_nobody_told_me_about_small_farming_i_cant_make_a_living/ (accessed 24 February 2021).

Muchow, R.C., Sinclair, T.R. and Bennett, J.M. (1990) Temperature and solar radiation effects on potential maize yield across locations. *Agronomy Journal* 82, 338–343.

Muller, A., Schader, C., El-Hage Scialabba, N., Buggemann, J., Isensee, A., *et al.* (2017) Strategies for feeding the world more sustainably with organic agriculture. *Nature Communications* 8, 1290.

Murata, Y. (1969) Physiological responses to nitrogen in plants. In: Eastin, J.D., Haskins, F.A., Sullivan, C.Y. and Van Bavel, C.H.M. (eds) *Physiological Aspects of Crop Yield.* American Society of Agronomy–Crop Science Society of America, Madison, Wisconsin, pp. 235–259.

NASS (2020) National Agricultural Statistics Service. Available at: https://www.nass.usda.gov (accessed 24 February 2021).

Nemali, K.S., Bonin, C., Doubleman, F.G., Stephens, M., Reeves, W.R., *et al.* (2014) Physiological responses related to grain yield under drought in the first biotechnology-derived drought-tolerant maize. *Plant, Cell & Environment* 38, 1866–1880.

NOAA (2016) Climate Data Online: Dataset Discovery. National Oceanic and Atmospheric Administration. Available at: http://ncdc.noaa.gov/cdo-web/datasets#Normal_Ann (accessed 27 August 2016).

O'Dowd, P. and Hagan, A. (2020) Singapore approved the sale of lab-grown chicken nuggets. One CEO hopes no-kill meat will go global. Available at: https://www.wbur.org/hereandnow/2020/12/11/just-lab-grown-chicken-meat (accessed 15 March 2021).

Okin, G.S. (2017) Environmental impacts of food consumption by dogs and cats. *PLoS One* 12(8), e0181301.

Olmstead, A.L. and Rhode, P.W. (2008) *Creating Abundance: Biological Innovation and American Agricultural Development.* Cambridge University Press, Cambridge.

Osterberg, J.T., Xiang, W., Olsen, L.I., Edenbrandt, A.K., Vedel, S.A., *et al.* (2017) Accelerating the domestication of new crops: feasibility and approaches. *Trends in Plant Science* 22, 373–384.

Penning de Vries, F.W.T., Brunsting, A.H.M. and van Laar, H.H. (1974) Products, requirements and efficiency of biosynthesis: a quantitative approach. *Journal of Theoretical Biology* 45, 339–377.

Pfeiffer, T.W. (1996) Choosing soybean varieties from yield trials: multiple maturity groups and yield variability. *Journal of Production Agriculture* 9, 371–376.

Pike, D.R., McGlamery, M.D. and Knake, E.L. (1991) A case study of herbicide use. *Weed Technology* 5, 639–646.

Ponisio, L.C., M'Gonigle, L.K., Mace, K.C., Palomino, J., de Valpine, P. and Krenen, C. (2015) Diversification practices reduce organic to conventional yield gap. *Proceedings of the Royal Society B: Biological Sciences* 282, 20141396.

Pszczola, D.E. (2012) Seeds of success. *Food Technology* 66, 45–55.

Rennie, J. (2020) Climate change 101. Sorting fact from fiction. *Scientific American* 29, 6–9.

Ritchie, S.W., Hanway, J.J. and Benson, G.O. (1993) *How a Corn Plant Develops.* Special Report No. 48. Iowa State University Cooperative Extension Service, Ames, Iowa.

Rosenberg, N.J., Blad, B.L. and Verma, S.B. (1983) *Microclimate: The Biological Environment.* Wiley, New York.

Ross, F., DiMatteo, J. and Gerrudo, A. (2020) Maize prolificacy: a source of reproductive plasticity that contributes to yield stability in drought prone environments. *Field Crops Research* 247, 107699.

Russell, W.A. (1991) Genetic improvement in maize yields. *Advances in Agronomy* 46, 245–298.

Schaffer, H.D. and Ray, D.E. (2019) Agricultural supply management and farm policy. *Renewable Agriculture and Food Systems* 35(Sp. Iss. 4), 453–462.

Schmitz, P.K., Stanley, J.D. and Kandel, H. (2020) Row spacing and seeding rate effect on soybean yield in North Dakota. *Crop, Forage & Turfgrass Management* 6(1), e20010.

Scott, H.D. (2000) *Soil Physics: Agriculture and Environmental Applications.* Iowa State University Press, Ames, Iowa.

Scott, W.O. and Aldrich, S.R. (1970) *Modern Soybean Production.* S & A Publications, Champaign, Illinois.

Shearman, V.J., Sylvester-Bradley, R., Scott, R.K. and Foulkes, M.J. (2005) Physiological processes associated with wheat yield progress in the UK. *Crop Science* 45, 175–185.

Shibles, R.M. and Weber, C.R. (1965) Leaf area, solar radiation interception, and dry matter production by soybean. *Crop Science* 5, 575–577.

Shimelis, H., Mashela, P.W. and Hugo, A. (2008) Performance of vernonia as an alternative industrial oil crop in Limpopo Province of South Africa. *Crop Science* 48, 236–242.

Sinclair, T.R. and de Wit, C.T. (1975) Comparative analysis of photosynthate and nitrogen requirements in the production of seeds by various crops. *Science* 18, 565–567.

Sinclair, T.R. and Muchow, R.C. (1999) Radiation use efficiency. *Advances in Agronomy* 65, 215–265.

Sinclair, T.R. and Sinclair, C.J. (2010) *Bread, Beer and the Seeds of Change: Agriculture's Imprint on World History.* CAB International, Wallingford, UK.

Smil, V. (2019) *Growth: From Microorganisms to Megacities.* The MIT Press, Cambridge, Massachusetts.

Spaeth, S.C. and Sinclair, T.R. (1984) Soybean seed growth. I. Timing of growth of individual seeds. *Agronomy Journal* 76, 123–127.

Spiegel, B. (2020) How David Hula grows 600-bushel-plus corn. Record-breaking corn yields take discipline, research, and resolve. *Successful Farming*, Crops News, 1 June 2020. Available at: https://www.agriculture.com/news/crops/how-david-hula-grows-600-bushel-plus-corn (accessed 15 March 2021).

Stanhill, G. (1976) Trends and deviations in the yield of the English wheat crop during the last 750 years. *Agro-Ecosystems* 3, 1–10.

TeKrony, D.M. and Egli, D.B. (1991) Relationship of seed vigor to crop yield: a review. *Crop Science* 31, 816–822.

TeKrony, D.M., Egli, D.B. and Henson, G. (1981) A visual indicator of physiological maturity in soybean plants. *Agronomy Journal* 73, 553–556.

Thomas, H. and Stoddart, J.L. (1980) Leaf senescence. *Annual Review of Plant Physiology* 31, 83–111.

Thompson, L.M. (1969) Weather and technology in the production of corn in the US corn belt. *Agronomy Journal* 61, 453–456.

Thompson, N.M., Bir, C., Wildmar, D.A. and Mintert, J.R. (2019) Farmer perceptions of precision agriculture benefits. *Journal of Agriculture and Applied Economics* 51, 142–163.

Thornthwaite, C.W. (1948) An approach toward a rational classification of climate. *Geographical Review* 38, 55–94.

Tilman, D.C., Blazer, C., Hill, I. and Befort, B.J. (2011) Global food demand and the sustainable intensification of agriculture. *Proceedings of the National Academy of Sciences USA* 108, 20260–20264.

Tollenaar, M. and Daynard, T.B. (1978) Kernel growth and development of two positions in the ear of maize (*Zea mays*). *Canadian Journal of Plant Science* 58, 189–197.

United Nations (2019) World Population Prospects 2019. Department of Economic and Social Affairs, Population Dynamics. Available at: https://population.un.org/wpp/ (accessed 23 January 2021).

Van Camp, W. (2005) Yield enhancement genes: seeds for growth. *Current Opinions in Biotechnology* 16, 147–153.

Vanderlip, R.L. and Reeves, H.I. (1972) Growth stages of sorghum (*Sorghum bicolor* Moench). *Agronomy Journal* 64, 13–16.

Vaughan, J.G. and Geissler, C.A. (1997) *The New Oxford Book of Food Plants*. Oxford University Press, Oxford.

Venkateswarlu, B., Rao, J.S. and Rao, A.V. (1977) Relationship between growth duration and yield parameters in irrigated rice (*Oryza sativa*). *Indian Journal of Plant Physiology* 20, 69–76.

Villalobos, F.J., Sadras, V.O., Soriano, A. and Fereres, E. (1994) Planting density effects on dry matter partitioning and productivity of sunflower hybrids. *Field Crops Research* 36, 1–11.

Viney, M. (ed.) (2005) *Micrometeorology in Agricultural Systems*. American Society of Agronomy–Crop Science Society of America–Soil Science Society of America, Madison, Wisconsin.

Wallace, H.A. and Brown, W.L. (1988) *Corn and Its Early Fathers*, revised edn. Iowa State University Press, Ames, Iowa.

Warren, J.M. (2015) *The Nature of Crops: How We Came to Eat the Plants We Do*. CAB International, Wallingford, UK.

Watson, D.J. (1947) Comparative physiological studies of the growth of field crops. II. Variation in net assimilation rate and leaf area between species and varieties and within and between years. *Annals of Botany* 11, 41–76.

Wells, R., Schulze, L.L., Ashley, D.A., Boerma, H.R. and Brown, R.H. (1982) Cultivar differences in canopy apparent photosynthesis and their relationship to seed yield in soybean. *Crop Science* 22, 886–890.

Wennblom, R.D. (1978) Have crop yields peaked out? *Farm Journal* 102, 32–34.

Wiggans, R.G. (1939) The influence of space and arrangement on the production of soybean plants. *Journal of the American Society of Agronomy* 31, 314–321.

Winter, S.R. and Ohlrogge, A.J. (1973) Leaf angle, leaf area, and corn (*Zea mays* L.) yield. *Agronomy Journal* 65, 395–397.

Wu, J., Lowit, S.J., Weers, B., Sun, J., Monagar, N., *et al.* (2019) Overexpression of *zmm28* increases maize grain yield in the field. *Proceedings of the National Academy of Sciences USA* 116, 23850–23858.

Yergin, D. (2020) *The New Map: Energy, Climate and the Clash of Nations.* Penguin Press, New York.

Zabinski, C. (2020) *Amber Waves: The Extraordinary Biography from Wild Grass to World Megacrop.* University of Chicago Press, Chicago, Illinois.

Zadoks, J.C., Chang, T.T. and Konzinks, C.F. (1974) A decimal code for the growth of cereals. *Weed Research* 14, 415–421.

Zeiher, C., Egli, D.B., Leggett, J.E. and Reicosky, D.A. (1982) Cultivar differences in N redistribution in soybeans. *Agronomy Journal* 74, 375–379.

Zhang, J., Fengier, K.A. and Van Hement, J.L. (2019) Identification and characterization of a novel stay-green QTL that increases yield in maize. *Plant Biotechnology Journal* 17, 2272–2285.

Zhang, O. (2007) Strategies for developing green super rice. *Proceedings of the National Academy of Sciences USA* 104, 16402–16409.

Index

CABI – who we are and what we do

This book is published by **CABI**, an international not-for-profit organisation that improves people's lives worldwide by providing information and applying scientific expertise to solve problems in agriculture and the environment.

CABI is also a global publisher producing key scientific publications, including world renowned databases, as well as compendia, books, ebooks and full text electronic resources. We publish content in a wide range of subject areas including: agriculture and crop science / animal and veterinary sciences / ecology and conservation / environmental science / horticulture and plant sciences / human health, food science and nutrition / international development / leisure and tourism.

The profits from CABI's publishing activities enable us to work with farming communities around the world, supporting them as they battle with poor soil, invasive species and pests and diseases, to improve their livelihoods and help provide food for an ever growing population.

CABI is an international intergovernmental organisation, and we gratefully acknowledge the core financial support from our member countries (and lead agencies) including:

UKaid
from the British people

Ministry of Agriculture
People's Republic of China

Agriculture and
Agri-Food Canada

Ministry of Foreign Affairs of the
Netherlands

Schweizerische Eidgenossenschaft
Confédération suisse
Confederazione Svizzera
Confederaziun svizra

Swiss Agency for Development
and Cooperation SDC

Discover more

To read more about CABI's work, please visit: **www.cabi.org**

Browse our books at: **www.cabi.org/bookshop**,
or explore our online products at: **www.cabi.org/publishing-products**

Interested in writing for CABI? Find our author guidelines here:
www.cabi.org/publishing-products/information-for-authors/